PDE Models for Atherosclerosis Computer Implementation in R

Synthesis Lectures on Mathematics and Statistics

Editor
Steven G. Krantz, *Washington University, St. Louis*

An Easy Path to Convex Analysis and Applications
Boris S. Mordukhovich and Nguyen Mau Nam
2013

Applications of Affine and Weyl Geometry
Eduardo García-Río, Peter Gilkey, Stana Nikčević, and Ramón Vázquez-Lorenzo
2013

Essentials of Applied Mathematics for Engineers and Scientists, Second Edition
Robert G. Watts
2012

Chaotic Maps: Dynamics, Fractals, and Rapid Fluctuations
Goong Chen and Yu Huang
2011

Matrices in Engineering Problems
Marvin J. Tobias
2011

The Integral: A Crux for Analysis
Steven G. Krantz
2011

Statistics is Easy! Second Edition
Dennis Shasha and Manda Wilson
2010

Lectures on Financial Mathematics: Discrete Asset Pricing
Greg Anderson and Alec N. Kercheval
2010

Jordan Canonical Form: Theory and Practice
Steven H. Weintraub
2009

The Geometry of Walker Manifolds
Miguel Brozos-Vázquez, Eduardo García-Río, Peter Gilkey, Stana Nikčević, and Ramón Vázquez-Lorenzo
2009

An Introduction to Multivariable Mathematics
Leon Simon
2008

Jordan Canonical Form: Application to Differential Equations
Steven H. Weintraub
2008

Statistics is Easy!
Dennis Shasha and Manda Wilson
2008

A Gyrovector Space Approach to Hyperbolic Geometry
Abraham Albert Ungar
2008

PDE Models for Atherosclerosis Computer Implementation in R
William E. Schiesser

ISBN: 978-3-031-01286-0   paperback
ISBN: 978-3-031-02414-6   ebook
ISBN: 978-3-031-00260-1   hardcover

DOI 10.1007/978-3-031-02414-6

A Publication in the Springer series
SYNTHESIS LECTURES ON MATHEMATICS AND STATISTICS

Lecture #22
Series Editor: Steven G. Krantz, *Washington University, St. Louis*
Series ISSN
Print 1938-1743   Electronic 1938-1751

PDE Models for Atherosclerosis Computer Implementation in R

William E. Schiesser
Lehigh University

SYNTHESIS LECTURES ON MATHEMATICS AND STATISTICS #22

ABSTRACT

Atherosclerosis is a pathological condition of the arteries in which plaque buildup and stiffening (hardening) can lead to stroke, myocardial infarction (heart attacks), and even death. Cholesterol in the blood is a key marker for atherosclerosis, with two forms: (1) LDL - low density lipoproteins and (2) HDL - high density lipoproteins. Low LDL and high HDL concentrations are generally considered essential for limited atherosclerosis and good health.

This book pertains to a mathematical model for the spatiotemporal distribution of LDL and HDL in the arterial endothelial inner layer (EIL, intima). The model consists of a system of six partial differential equations (PDEs) with the dependent variables

1. $\ell(x, t)$: concentration of modified LDL

2. $h(x, t)$: concentration of HDL

3. $p(x, t)$: concentration of chemoattractants

4. $q(x, t)$: concentration of ES cytokines

5. $m(x, t)$: density of monocytes/macrophages

6. $N(x, t)$: density of foam cells

   and independent variables

1. x: distance from the inner arterial wall

2. t: time

The focus of this book is a discussion of the methodology for placing the model on modest computers for study of the numerical solutions. The foam cell density $N(x, t)$ as a function of the bloodstream LDL and HDL concentrations is of particular interest as a precursor for arterial plaque formation and stiffening.

The numerical algorithm for the solution of the model PDEs is the method of lines (MOL), a general procedure for the computer-based numerical solution of PDEs. The MOL coding (programming) is in R, a quality, open-source scientific computing system that is readily available from the Internet. The R routines for the PDE model are discussed in detail, and are available from a download link so that the reader/analyst/researcher can execute the model to duplicate the solutions reported in the book, then experiment with the model, for example, by changing the parameters (constants) and extending the model with additional equations.

KEYWORDS

atherosclerosis, cholesterol, LDL, HDL, partial differential equations, method of lines

Contents

Preface

Atherosclerosis is a pathological condition of the arteries in which plaque buildup and stiffening (hardening) can lead to stroke, myocardial infarction (heart attacks), and even death. Cholesterol in the blood is a key marker for atherosclerosis, with two forms: (1) LDL - low density lipoproteins and (2) HDL - high density lipoproteins. Low LDL and high HDL concentrations are generally considered essential for limited atherosclerosis and good health.

This book pertains to a mathematical model for the spatiotemporal distribution of LDL and HDL in the arterial endothelial inner layer (EIL, intima). The model consists of a system of six partial differential equations (PDEs) with the dependent variables

1. $\ell(x, t)$: concentration of modified LDL

2. $h(x, t)$: concentration of HDL

3. $p(x, t)$: concentration of chemoattractants

4. $q(x, t)$: concentration of ES cytokines

5. $m(x, t)$: density of monocytes/macrophages

6. $N(x, t)$: density of foam cells

and independent variables

1. x: distance from the inner arterial wall

2. t: time

These variables are discussed in detail in [1–4]. Authoritative introductions to atherosclerosis are given with Google searches for: (1) NIH atherosclerosis, (2) MGH atherosclerosis and (3) Mayo atherosclerosis. The emphasis of this book is a discussion of the methodology for placing the model on modest computers for study of the numerical solutions, specifically, the six dependent variables $\ell(x, t)$ to $N(x, t)$ as a function of x and t. The foam cell density $N(x, t)$ as a function of the bloodstream LDL and HDL concentrations is of particular interest as a precursor for arterial plaque formation and stiffening.

The numerical algorithm for the solution of the model PDEs is the method of lines (MOL), a general procedure for the computer-based numerical solution of PDEs. The MOL coding (programming) is in R, a quality, open-source scientific computing system that is readily available from the Internet. The R routines for the PDE model are discussed in detail, and are available from a download link so that the reader/analyst/researcher can execute the model to duplicate the solutions reported in the book, then experiment with the model, for example, by changing the parameters (constants) and extending the model with additional equations.

In summary, the intent of the book is to provide variants of a computer-based model that can be used to study the spatiotemporal dynamics of atherosclerosis. The cases discussed in the book include the possible use of LDL-lowering and HDL-raising drug therapy for altering foam cell production, and thereby, arterial plaque formation and stiffening.

I would welcome comments about the application and usefulness of the models and their computer implementation.

William E. Schiesser
October 2018

REFERENCES

[1] Chalmers, A. D., Cohen, A., Bursill, C. A., and Myerscough, M. R. (2015), Bifurcation and dynamics in a mathematical model of early atherosclerosis, *Journal of Mathematical Biology*, 71, pp. 1451–1480. DOI: 10.1007/s00285-015-0864-5. xi

[2] Chalmers, A. D., Bursill, C. A., and Myerscough, M. R. (2017), Nonlinear dynamics of early atherosclerosis plaque formation may determine the effect of high density lipoproteins in plaque regressions, *Plos One*, 12, no. 11, November. DOI: 10.1371/journal.pone.0187674.

[3] Hao, W. and Friedman, A. (2014), The LDL-HDL profile determines the risk of atherosclerosis: A mathematical model, *Plos One*, 9, no. 3, March. DOI: 10.1371/journal.pone.0090497.

[4] Khatib, N. E., Genieys, S., Kazmierczak, B., and Volpert, V. (2012), Reaction-diffusion model of atherosclerosis development, *Journal of Mathematical Biology*, 65, pp. 349–374. DOI: 10.1007/s00285-011-0461-1. xi

CHAPTER 1

PDE Model Formulation

1.1   INTRODUCTION

A mathematical model is presented in this book that includes basic molecular and cellular mechanisms for the development of atherosclerosis. The model is formulated as a set of partial differential equations (PDEs) that define the space and time (spatiotemporal) variation of the concentration or densities of six dependent variables (see Table 1.1).

Table 1.1: Dependent variables of the PDE model [1, 2]

$\ell(x,t)$	concentration of modified LDL
$h(x,t)$	concentration of HDL
$p(x,t)$	concentration of chemoattractants
$q(x,t)$	concentration of ES cytokines
$m(x,t)$	density of monocytes/macrophages
$N(x,t)$	density of foam cells

A basic schematic diagram for the model is given in Fig. 1.1.

The RHS terms of Eqs. (1.1) are identified in Fig. 1.1 with the corresponding coefficients, e.g., D_ℓ for $D_\ell \dfrac{\partial^2 \ell}{\partial x^2}$, $-r_{\ell 1}$ for $-r_{\ell 1} \dfrac{\ell}{r_{l3} + \ell} m$. The second example also illustrates the nonlinear coupling between the PDEs, (ℓ from Eq. (1.1-1) and m from Eq. (1.1-5)), but this coupling is not included in Fig. 1.1. A minus sign with the coefficients in Fig. 1.1 indicates the corresponding term is for depletion and therefore reduces the corresponding PDE LHS derivative in t. More detailed diagrams of the model with the PDE coupling indicated are given in [1, 2].

The PDEs have one dimension (1D), the distance x across the inner layer of the arterial wall, that is, the endothelium inner layer (EIL) or intima. The model solutions give the variation of the six dependent variables with respect to x and time t in the EIL. The model PDEs are programmed in R, an open-source scientific programming system that is readily available as a download from the Internet.

A principal output of the model is the spatiotemporal distribution of foam cells which lead to arterial stiffness and plaque that contribute to atherosclerosis pathology.

The model responds to the concentration of low and high density lipoproteins (LDL, HDL) in the bloodstream (lumen). The model output is the time evolution of the six PDE de-

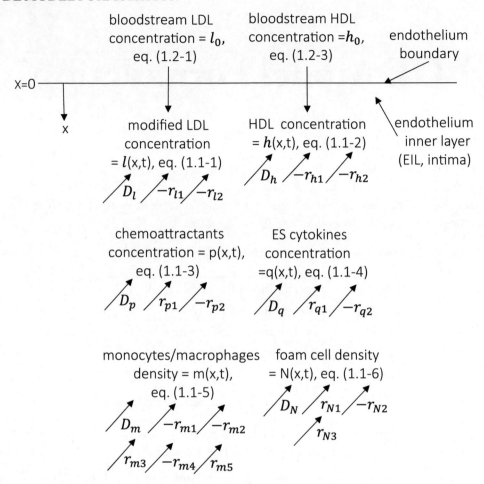

Figure 1.1: Diagram of the PDE model, Eqs. (1.1).

pendent variables from homogeneous (zero) initial conditions (ICs) in numerical and graphical format.

The model is intended as a prototype that the reader/analyst/researcher can study by using the R routines provided with the book. The programmed base case can first be executed to confirm the numerical and graphical output discussed in the book. The model can then be modified through variation of the parameters, and extended through the modification of and addition to the PDEs. The basic numerical algorithm is the numerical method of lines (MOL) in which the PDEs are approximated as a system of initial value ordinary differential equations (ODEs) that can be integrated (solved) numerically with a library ODE integrator [5, 6].

As is usually the case, the model has several parameters that are given nominal values in the main program of the R routines, but are subject to improved estimation/refinement based on reported new insights and experimental data. The numerical values of the parameters were selected to give a time scale of two months, and observed diffusion rates across the EIL (since the endothelial layer is thin, e.g., 40 μm,[1] the variation in x is small).

The formulation of the model is discussed next.

1.2 PDE MODEL

The six PDEs are generally based on conservation principles, with spatial transport according to Fick's first law for diffusion. For example, for the modified LDL concentration,[2] $\ell(x, t)$ a mass balance on an incremental volume $A_c \Delta x$ gives

$$A_c \Delta x \frac{\partial \ell}{\partial t} = A_c q_x|_x - A_c q_x|_{x+\Delta x} - A_c \Delta x r_{\ell 1} \frac{\ell}{r_{l3} + \ell} m - A_c \Delta x r_{\ell 2} \ell,$$

A_c is an area that is normal (perpendicular) to the mass transfer. Δx is an incremental length in x. The net diffusion of modified LDL into or out of the incremental volume $A_c \Delta x$ is $A_c q_x|_x - A_c q_x|_{x+\Delta x}$. The physical significance of the other terms in this balance is discussed below.

The diffusion flux of modified LDL is given by Fick's first law.

$$q_x = -D_\ell \frac{\partial \ell}{\partial x}.$$

The minus is included so that the flux is in the direction of decreasing ℓ (the gradient $\frac{\partial \ell}{\partial x}$ is negative).

Substitution of the flux in the mass balance gives

$$\frac{\partial \ell}{\partial t} = D_\ell \frac{\frac{\partial \ell}{\partial x}|_{x+\Delta x} - \frac{\partial \ell}{\partial x}|_x}{\Delta x} - r_{\ell 1} \frac{\ell}{r_{l3} + \ell} m - r_{\ell 2} \ell.$$

In the limit $\Delta x \to 0$, a partial derivative $\frac{\partial^2 \ell}{\partial x^2}$ results and the ℓ balance is

$$\frac{\partial \ell}{\partial t} = D_\ell \frac{\partial^2 \ell}{\partial x^2} - r_{\ell 1} \frac{\ell}{r_{l3} + \ell} m - r_{\ell 2} \ell. \tag{1.1-1}$$

Equation (1.1-1) is the first PDE in the system of six PDEs. The terms in Eq. (1.1-1) apply to the interior of the EIL adjacent to the bloodstream [1].

---

[1] 1 *micron* (μm) = 1 *micrometer* = 10^{-6} m = 10^{-4} cm = 10^3 *nanometers* (*nm*). The model is expressed in cgs units.
[2] The LDL in the blood stream when transported to the EIL is modified, for example, by oxidation. It is therefore termed *modified LDL*, and its concentration is designated as ℓ.

$D_\ell \dfrac{\partial^2 \ell}{\partial x^2}$: Net diffusion of modified LDL.

$-r_{\ell 1} \dfrac{\ell}{r_{l3} + \ell} m$: Loss of modified LDL from consumption by macrophages. This rate is linear in macrophage density, m, but saturating in modified LDL concentration, ℓ, according to Michaelis–Menten kinetics (representing the limiting rate of consumption of modified LDL by each macrophage [1]).

$-r_{\ell 2}\ell$: Linear decay term which represents other losses of modified LDL, e.g., through degradation.

$\dfrac{\partial \ell}{\partial t}$: Accumulation (or depletion if negative) of modified LDL resulting from the sum of the preceding RHS terms. This derivative determines how ℓ evolves in time t.

Similarly, a mass balance on the HDL with concentration $h(x, t)$ gives

$$\frac{\partial h}{\partial t} = D_h \frac{\partial^2 h}{\partial x^2} - r_{h1} \frac{h}{r_{h3} + h} N - r_{h2} h. \tag{1.1-2}$$

The terms in Eq. (1.1-2) again apply to the EIL adjacent to the bloodstream [1].

$D_h \dfrac{\partial^2 h}{\partial x^2}$: Diffusion of HDL.

$-r_{h1} \dfrac{h}{r_{h3} + h} N$: Loss of HDL particles which have the capacity to take up cholesterol from foam cells via reverse cholesterol transport (RCT) and thereby limit the increase in plaque. This term saturates as h increases, since each foam cell has a finite number of transporters in its membrane and so there is a limit to the rate that each cell can engage with HDL particles ([2, Fig. 1]).

$-r_{h2}h$: Linear decay term which represents other losses of HDL, e.g., through degradation.

$\dfrac{\partial h}{\partial t}$: Accumulation (or depletion if negative) of HDL resulting from the sum of the preceding RHS terms. This derivative determines how h evolves in time t.

A mass balance for the chemoattractants with concentration $p(x, t)$ gives

$$\frac{\partial p}{\partial t} = D_p \frac{\partial^2 p}{\partial x^2} + r_{p1} \frac{\ell}{r_{l3} + \ell} m - r_{p2} p. \tag{1.1-3}$$

The terms in Eq. (1.1-3) pertain to the variation of the concentration of monocyte chemoattractants, $p(x, t)$, in the interior of the intima.

$D_p \dfrac{\partial^2 p}{\partial x^2}$: Diffusion of monocyte chemoattractants.

$r_{p1} \dfrac{\ell}{r_{l3} + \ell} m$: Increase (production) of chemoattractants by macrophages at a rate proportional to the macrophage consumption of modified LDL.

$-r_{p2} p$: Linear decay term which models other losses of chemoattractants.

$\dfrac{\partial p}{\partial t}$: Accumulation (or depletion if negative) of chemoattractants resulting from the sum of the preceding RHS terms. This derivative determines how p evolves in time t.

A mass balance for the ES cytokines with concentration $q(x, t)$ gives

$$\frac{\partial q}{\partial t} = D_q \frac{\partial^2 q}{\partial x^2} + r_{q1} \frac{\ell}{r_{l3} + \ell} m - r_{q2} q. \tag{1.1-4}$$

The terms in Eq. (1.1-4) pertain to the variation of the concentration of ES cytokines, $q(x, t)$, in the interior of the intima.

$D_q \dfrac{\partial^2 q}{\partial x^2}$: Diffusion of ES cytokines.

$+r_{q1} \dfrac{\ell}{r_{l3} + \ell} m$: Increase (production) of ES cytokines by macrophages at a rate proportional to the macrophage consumption of modified LDL.

$-r_{q2} q$: Linear decay term which models generic ES cytokine loss in the intima.

$\dfrac{\partial q}{\partial t}$: Accumulation (or depletion if negative) of ES cytokines resulting from the sum of the preceding RHS terms. This derivative determines how q evolves in time t.

A mass balance for the macrophages with concentration $m(x, t)$ includes chemotaxis diffusion. The net diffusion flux is

$$q_x = r_{m1} m \frac{\partial \ell}{\partial x}\big|_x - \left(r_{m1} m \frac{\partial \ell}{\partial x}\big|_{x+\Delta x} \right).$$

The diffusivity r_{m1} is a constant, but it multiplies $m(x, t)$ so that this flux is nonlinear, with the derivative (gradient) $\dfrac{\partial \ell}{\partial x}$.

A finite difference (FD) is again formed, and with $\Delta x \to 0$, a partial derivative results, which can be expanded by differentiating the product $m \dfrac{\partial \ell}{\partial x}$.

$$-r_{m1} \lim_{\Delta x \to 0} \frac{m \frac{\partial \ell}{\partial x}\big|_{x+\Delta x} - m \frac{\partial \ell}{\partial x}\big|_x}{\Delta x} =$$

$$-r_{m1} \frac{\partial}{\partial x}(m \frac{\partial \ell}{\partial x}) = -r_{m1} \left(\frac{\partial m}{\partial x} \frac{\partial \ell}{\partial x} + m \frac{\partial^2 \ell}{\partial x^2} \right).$$

This expanded form of the chemotaxis diffusion term is then used in the mass balance for $m(x,t)$.

$$\frac{\partial m}{\partial t} = D_m \frac{\partial^2 m}{\partial x^2} - r_{m1}(\frac{\partial m}{\partial x} \frac{\partial l}{\partial x} + m \frac{\partial^2 l}{\partial x^2})$$
$$- r_{m2} \frac{\ell}{r_{l3} + \ell} m + r_{m3} \frac{h}{r_{h3} + h} N - r_{m4}m + r_{m5}\ell. \qquad (1.1\text{-}5)$$

The terms in Eq. (1.1-5) pertain to the variation of the density of macrophages, $m(x,t)$, in the EIL.

$D_q \dfrac{\partial^2 m}{\partial x^2}$: The linear diffusion of macrophages.

$-r_{m1} \dfrac{\partial}{\partial x} \left(m \dfrac{\partial \ell}{\partial x} \right) = -r_{m1} \left(\dfrac{\partial m}{\partial x} \dfrac{\partial l}{\partial x} + m \dfrac{\partial^2 l}{\partial x^2} \right)$: The nonlinear chemotactic diffusion.

$-r_{m2} \dfrac{\ell}{r_{l3} + \ell} m$: Rate of conversion of macrophages to foam cells in the EIL after the consumption of modified LDL.

$+r_{m3} \dfrac{h}{r_{h3} + h} N$: Rate that foam cells undergo reverse cholesterol transport (RCT) and return to the pool of inflammatory macrophages.

$-r_{m4}m$: Small linear decay term that accounts for other losses and cell death and, perhaps, differentiation.

$+r_{m5}\ell$: Rate of increase of macrophages from modified LDL.[3]

$\dfrac{\partial m}{\partial t}$: Accumulation (or depletion if negative) of macrophages resulting from the sum of the preceding RHS terms. This derivative determines how m evolves in time t.

A mass balance for the foam cells with concentration $N(x,t)$ gives

$$\frac{\partial N}{\partial t} = D_N \frac{\partial^2 N}{\partial x^2} + r_{N1} \frac{\ell}{r_{l3} + \ell} m - r_{N2} \frac{h}{r_{h3} + h} N + r_{N3}\ell. \qquad (1.1\text{-}6)$$

The terms in Eq. (1.1-5) pertain to the variation of the density of foam cells, $N(x,t)$, in the EIL.

[3] The terms $r_{m5}\ell$ and $r_{N3}\ell$ were added by the author (WES) to Eqs. (1.1-5) and (1.1-6), respectively, to give nonzero RHS terms. Otherwise, all of the RHS terms in Eqs. (1.1-5) and (1.1-6) are zero initially and remain at zero so the solutions $m(x,t), N(x,t)$, as well as $p(x,t), q(x.t)$, do not move away from the homogeneous (zero) ICs of Eqs. (1.3). This special case is demonstrated subsequently.

$D_N \dfrac{\partial^2 N}{\partial x^2}$: Diffusion of foam cells.

$+r_{N1}\dfrac{\ell}{r_{l3} + \ell}m$: Rate of conversion of macrophages to foam cells in the intima after the consumption of modified LDL.

$-r_{N2}\dfrac{h}{r_{h3} + h}N$: Rate that foam cells undergo RCT and return to the pool of inflammatory macrophages.

$+r_{N3}\ell$: Rate of increase of foam cells from modified LDL.[3]

$\dfrac{\partial N}{\partial t}$: Accumulation (or depletion if negative) of macrophages resulting from the sum of the preceding RHS terms. This derivative determines how N evolves in time t.

This completes the derivation of the six PDEs, Eqs. (1.1). Each of these PDEs is first order in t and second order in x. Therefore, each requires one IC and two BCs. The BCs follow as Eqs. (1.2-1) to (1.2-12).[4]

$$D_\ell \frac{\partial \ell(x = x_l, t)}{\partial x} = -k_l(\ell_0 - \ell(x = x_l, t)); \qquad \text{(1.2-1)}$$

$$\frac{\partial \ell(x = x_u, t)}{\partial x} = 0 \qquad \text{(1.2-2)}$$

$$D_h \frac{\partial h(x = x_l, t)}{\partial x} = -k_h(h_0 - h(x = x_l, t)); \qquad \text{(1.2-3)}$$

$$\frac{\partial h(x = x_u, t)}{\partial x} = 0 \qquad \text{(1.2-4)}$$

$$D_p \frac{\partial p(x = x_l, t)}{\partial x} = -k_p(p_0 - p(x = x_l, t)); \qquad \text{(1.2-5)}$$

$$\frac{\partial p(x = x_u, t)}{\partial x} = 0 \qquad \text{(1.2-6)}$$

[4]Conventional Robin BCs based on continuity of mass transfer rates across the EIL are specified first for each PDE. The mass transfer rates across this surface can be adjusted through the selection of coefficients k_l to k_N as demonstrated subsequently.

The BCs stated in [1, 2] can be used based on the boundary concentrations l[1] to N[1]. Attempts to do this produced unrealistic/unacceptable results, e.g., negative and/or large concentrations. If the reader/analyst/researcher implements these surface BCs, additional parameters will have to be specified.

$$D_q \frac{\partial q(x = x_l, t)}{\partial x} = -k_q(q_0 - q(x = x_l, t)); \qquad (1.2\text{-}7)$$

$$\frac{\partial q(x = x_u, t)}{\partial x} = 0 \qquad (1.2\text{-}8)$$

$$D_m \frac{\partial m(x = x_l, t)}{\partial x} = -k_m(m_0 - m(x = x_l, t)); \qquad (1.2\text{-}9)$$

$$\frac{\partial m(x = x_u, t)}{\partial x} = 0 \qquad (1.2\text{-}10)$$

$$D_N \frac{\partial N(x = x_l, t)}{\partial x} = -k_N(N_0 - N(x = x_l, t)); \qquad (1.2\text{-}11)$$

$$\frac{\partial N(x = x_u, t)}{\partial x} = 0. \qquad (1.2\text{-}12)$$

The ICs for Eqs. (1.1) follow.

$$\ell(x, t = 0) = f_\ell(x) \qquad (1.3\text{-}1)$$

$$h(x, t = 0) = f_h(x) \qquad (1.3\text{-}2)$$

$$p(x, t = 0) = f_p(x) \qquad (1.3\text{-}3)$$

$$q(x, t = 0) = f_q(x) \qquad (1.3\text{-}4)$$

$$m(x, t = 0) = f_m(x) \qquad (1.3\text{-}5)$$

$$N(x, t = 0) = f_N(x). \qquad (1.3\text{-}6)$$

f_h to f_N are functions to be specified. For the subsequent discussion, they are taken as the zero function.

1.3 SUMMARY AND CONCLUSIONS

Equations (1.1), (1.2), and (1.3) constitute the PDE model, which is implemented in the next Chapter 2. The R coding (programming) of the MOL is discussed in detail, and the output from the R routines is analyzed in the subsequent Chapter 3.

The intent of the model is to provide a quantitative framework for the investigation of atherosclerosis on modest computers. Hopefully it will provide a small step toward an improved understanding of atherosclerosis. The basic concepts of the molecular and cellular mechanisms represented by the PDE model are discussed in detail in [1–4], and these references are gratefully acknowledged.

REFERENCES

[1] Chalmers, A. D., Cohen, A., Bursill, C. A., and Myerscough, M. R. (2015), Bifurcation and dynamics in a mathematical model of early atherosclerosis, *Journal of Mathematical Biology*, 71, pp. 1451–1480. DOI: 10.1007/s00285-015-0864-5. 1, 3, 4, 7, 9

[2] Chalmers, A. D., Bursill, C. A., and Myerscough, M. R. (2017), Nonlinear dynamics of early atherosclerosis plaque formation may determine the effect of high density lipoproteins in plaque regressions, *Plos One*, 12, no. 11, November. DOI: 10.1371/journal.pone.0187674. 1, 4, 7

[3] Hao, W. and Friedman, A. (2014), The LDL-HDL profile determines the risk of atherosclerosis: A mathematical model, *Plos One*, 9, no. 3, March. DOI: 10.1371/journal.pone.0090497.

[4] Khatib, N. E., Genieys, S., Kazmierczak, B., and Volpert, V. (2012), Reaction-diffusion model of atherosclerosis development, *Journal of Mathematical Biology*, 65, pp. 349–374. DOI: 10.1007/s00285-011-0461-1. 9

[5] Schiesser, W. E. and Griffiths, G. W. (2009), *A Compendium of Partial Differential Equation Models*, Cambridge University Press, Cambridge, UK. DOI: 10.1017/cbo9780511576270. 2

[6] Soetaert, K., Cash, J., and Mazzia, F. (2012), *Solving Differential Equations in R*, Springer-Verlag, Heidelberg, Germany. DOI: 10.1007/978-3-642-28070-2. 2

CHAPTER 2

PDE Model Implementation

2.1 INTRODUCTION

The implementation of the PDE model for atherosclerosis formulated in Chapter 1 is developed in this chapter.

2.2 PDE MODEL ROUTINES

The coding of the PDE model, Eqs. (1.1), (1.2), (1.3), starts with the main program.

2.2.1 MAIN PROGRAM

Listing 2.1: Main program for Eqs. (1.1), (1.2), (1.3)

```
#
# Six PDE atherosclerosis model
#
# Delete previous workspaces
  rm(list=ls(all=TRUE))
#
# Access ODE integrator
  library("deSolve");
#
# Access functions for numerical solution
  setwd("f:/atherosclerosis/chap2");
  source("pde1a.R");
#
# Select plotting
#
#   ip=1: 2D, vs x, t parametrically
#
#   ip=2: 2D vs t, x=xl
#
#   ip=3: 3D
#
```

```
  ip=3;
#
# Parameters
#
# Diffusivities
  D_l=1.0e-08; D_h=1.0e-08;
  D_p=1.0e-08; D_q=1.0e-08;
  D_m=1.0e-08; D_N=1.0e-08;
#
# Mass transfer coefficients
  k_l=1.0e-08; k_h=1.0e-08;
  k_p=0.0e-08; k_q=1.0e-08;
  k_m=0.0e-08; k_N=0.0e-08;
#
# Bloodstream (lumen) concentrations
  l0=1; h0=3;
  p0=0; q0=0;
  m0=0; N0=0;
#
# Reaction kinetic rate constants
#
# l(x,t)
  r_l1=5.0e-08; r_l2=5.0e-08;
  r_l3=1.0e-05;
#
# h(x,t);
  r_h1=5.0e-08; r_h2=5.0e-08;
  r_h3=1.0e-05;
#
# p(x,t)
  r_p1=5.0e-08; r_p2=5.0e-08;
#
# q(x,t)
  r_q1=5.0e-08; r_q2=5.0e-08;
#
# m(x,t)
  r_m1=5.0e-08; r_m2=5.0e-08;
  r_m3=5.0e-08; r_m4=5.0e-08;
  r_m5=5.0e-08;
```

```
#
# N(x,t)
  r_N1=5.0e-08; r_N2=5.0e-08;
  r_N3=5.0e-08;
#
# Spatial grid (in x)
  nx=26;
  xl=0;xu=0.004;
  x=seq(from=xl,to=xu,(xu-xl)/(nx-1));
#
# Independent variable for ODE integration
  t0=0;tf=1.0e+07;nout=51;
  tout=seq(from=t0,to=tf,by=(tf-t0)/(nout-1));
#
# Initial condition (t=0)
  u0=rep(0,6*nx);
  for(i in 1:nx){
    u0[i]     =0;
    u0[i+nx]  =0;
    u0[i+2*nx]=0;
    u0[i+3*nx]=0;
    u0[i+4*nx]=0;
    u0[i+5*nx]=0;
  }
  ncall=0;
#
# ODE integration
  out=lsodes(y=u0,times=tout,func=pde1a,
      sparsetype="sparseint",rtol=1e-6,
      atol=1e-6,maxord=5);
  nrow(out)
  ncol(out)
#
# Arrays for plotting numerical solution
  l=matrix(0,nrow=nx,ncol=nout);
  h=matrix(0,nrow=nx,ncol=nout);
  p=matrix(0,nrow=nx,ncol=nout);
  q=matrix(0,nrow=nx,ncol=nout);
  m=matrix(0,nrow=nx,ncol=nout);
```

```
  N=matrix(0,nrow=nx,ncol=nout);
  for(it in 1:nout){
    for(i in 1:nx){
      l[i,it]=out[it,i+1];
      h[i,it]=out[it,i+1+nx];
      p[i,it]=out[it,i+1+2*nx];
      q[i,it]=out[it,i+1+3*nx];
      m[i,it]=out[it,i+1+4*nx];
      N[i,it]=out[it,i+1+5*nx];
    }
  }
#
# Display numerical solution
  iv=seq(from=1,to=nout,by=25);
  for(it in iv){
    cat(sprintf(
      "\n                       t               x"));
    cat(sprintf(
      "\n                 l(x,t)          h(x,t)"));
    cat(sprintf(
      "\n                 p(x,t)          q(x,t)"));
    cat(sprintf(
      "\n                 m(x,t)          N(x,t)"));
  iv=seq(from=1,to=nx,by=25);
  for(i in iv){
    cat(sprintf("\n         %12.2e %12.2e",
                tout[it],x[i]));
    cat(sprintf("\n         %12.3e %12.3e",
                l[i,it],h[i,it]));
    cat(sprintf("\n         %12.3e %12.3e",
                p[i,it],q[i,it]));
    cat(sprintf("\n         %12.3e %12.3e\n",
                m[i,it],N[i,it]));
    }
  }
#
# Calls to ODE routine
  cat(sprintf("\n\n ncall = %5d\n\n",ncall));
#
```

```
# Plot PDE solutions
#
# l,h,p,q,m,N against x, parametrically in t
#
  if(ip==1){
#
# l(x,t)
  par(mfrow=c(1,1));
  matplot(x,l,type="l",xlab="x",ylab="l(x,t)",
    lty=1,main="",lwd=2,col="black");
#
# h(x,t)
  par(mfrow=c(1,1));
  matplot(x,h,type="l",xlab="x",ylab="h(x,t)",
    lty=1,main="",lwd=2,col="black");
#
# p(x,t)
  par(mfrow=c(1,1));
  matplot(x,p,type="l",xlab="x",ylab="p(x,t)",
    lty=1,main="",lwd=2,col="black");
#
# q(x,t)
  par(mfrow=c(1,1));
  matplot(x,q,type="l",xlab="x",ylab="q(x,t)",
    lty=1,main="",lwd=2,col="black");
#
# m(x,t)
  par(mfrow=c(1,1));
  matplot(x,m,type="l",xlab="x",ylab="m(x,t)",
    lty=1,main="",lwd=2,col="black");
#
# N(x,t)
  par(mfrow=c(1,1));
  matplot(x,N,type="l",xlab="x",ylab="N(x,t)",
    lty=1,main="",lwd=2,col="black");
  }
#
# l,h,p,q,m,N against t, x = xl
#
```

```
  if(ip==2){
#
# l(x,t)
  par(mfrow=c(1,1));
  matplot(tout,l[1,],type="l",xlab="t",
    ylab="l(x,t)",lty=1,main="",lwd=2,
    col="black");
#
# h(x,t)
  par(mfrow=c(1,1));
  matplot(tout,h[1,],type="l",xlab="t",
    ylab="h(x,t)",lty=1,main="",lwd=2,
    col="black");
#
# p(x,t)
  par(mfrow=c(1,1));
  matplot(tout,p[1,],type="l",xlab="t",
    ylab="p(x,t)",lty=1,main="",lwd=2,
    col="black");
#
# q(x,t)
  par(mfrow=c(1,1));
  matplot(tout,q[1,],type="l",xlab="t",
    ylab="q(x,t)",lty=1,main="",lwd=2,
    col="black");
#
# m(x,t)
  par(mfrow=c(1,1));
  matplot(tout,m[1,],type="l",xlab="t",
    ylab="m(x,t)",lty=1,main="",lwd=2,
    col="black");
#
# N(x,t)
  par(mfrow=c(1,1));
  matplot(tout,N[1,],type="l",xlab="t",
    ylab="N(x,t)",lty=1,main="",lwd=2,
    col="black");
  }
#
```

```
#  l,h,p,q,m,N,  3D
#
  if(ip==3){
  t=tout
#
# l(x,t)
  persp(x,t,l,theta=60,phi=45,
        xlim=c(xl,xu),ylim=c(t0,tf),xlab="x",
        ylab="t",zlab="l(x,t)");
#
# h(x,t)
  persp(x,t,h,theta=60,phi=45,
        xlim=c(xl,xu),ylim=c(t0,tf),xlab="x",
        ylab="t",zlab="h(x,t)");
#
# p(x,t)
  persp(x,t,p,theta=60,phi=45,
        xlim=c(xl,xu),ylim=c(t0,tf),xlab="x",
        ylab="t",zlab="p(x,t)");
#
# q(x,t)
  persp(x,t,q,theta=60,phi=45,
        xlim=c(xl,xu),ylim=c(t0,tf),xlab="x",
        ylab="t",zlab="q(x,t)");
#
# m(x,t)
  persp(x,t,m,theta=60,phi=45,
        xlim=c(xl,xu),ylim=c(t0,tf),xlab="x",
        ylab="t",zlab="m(x,t)");
#
# N(x,t)
  persp(x,t,N,theta=60,phi=45,
        xlim=c(xl,xu),ylim=c(t0,tf),xlab="x",
        ylab="t",zlab="N(x,t)");
  }
```

We can note the following details about Listing 2.1.

• Previous workspaces are deleted.

```
#
# Six PDE atherosclerosis model
#
# Delete previous workspaces
  rm(list=ls(all=TRUE))
```

- The R ODE integrator library deSolve is accessed. Then the directory with the files for the solution of Eqs. (1.1), (1.2), (1.3) is designated. Note that setwd (set working directory) uses / rather than the usual \.

```
#
# Access ODE integrator
  library("deSolve");
#
# Access functions for numerical solution
  setwd("f:/atherosclerosis");
  source("pde1a.R");
```

pde1a.R is the routine for the method of lines (MOL) approximation of PDEs (1.1) (discussed subsequently).

- An option for plotting is selected with ip.

```
#
# Select plotting
#
#   ip=1: 2D, vs x, t parametrically
#
#   ip=2: 2D vs t, x=xl
#
#   ip=3: 3D
#
    ip=3;
```

- The model parameters are defined numerically.

```
#
# Parameters
#
# Diffusivities
  D_l=1.0e-08; D_h=1.0e-08;
```

```
   D_p=1.0e-08; D_q=1.0e-08;
   D_m=1.0e-08; D_N=1.0e-08;
#
# Mass transfer coefficients
   k_l=1.0e-08; k_h=1.0e-08;
   k_p=0.0e-08; k_q=1.0e-08;
   k_m=0.0e-08; k_N=0.0e-08;
#
# Bloodstream (lumen) concentrations
   l0=1; h0=3;
   p0=0; q0=0;
   m0=0; N0=0;
#
# Reaction kinetic rate constants
#
# l(x,t)
   r_l1=1.0e-08; r_l2=1.0e-08;
   r_l3=1.0e-05;
#
# h(x,t);
   r_h1=1.0e-08; r_h2=1.0e-08;
   r_h3=1.0e-05;
#
# p(x,t)
   r_p1=1.0e-08; r_p2=1.0e-08;
#
# q(x,t)
   r_q1=1.0e-08; r_q2=1.0e-08;
#
# m(x,t)
   r_m1=1.0e-08; r_m2=1.0e-08;
   r_m3=1.0e-08; r_m4=1.0e-08;
   r_m5=1.0e-08;
#
# N(x,t)
   r_N1=1.0e-08; r_N2=1.0e-08;
   r_N3=1.0e-08;
```

These parameters require some additonal explanation.

- The diffusivities, D_ℓ for Eq. (1.1-1) to D_N for Eq. (1.1-6), are defined numerically with the units cm^2/s (1×10^{-8} for large molecules in liquids).

- The mass transfer coefficients, k_ℓ for Eq. (1.2-1) to k_N for Eq. (1.2-11), are defined numerically with the units cm/s. They have nonzero values for $\ell(x, t), h(x, t), q(x, t)$ in BCs (2.2) representing transfer to/from the bloodstream.

- The bloodstream concentrations, ℓ_0 for Eq. (1.2-1) to N_0 for Eq. (1.2-11), are defined numerically with normalized values. They are nonzero for $\ell(x, t), h(x, t)$ for LDL, HDL, respectively, in the bloodstream (HDL is three times LDL, a typical ratio for normal (healthy) blood lipoprotein concentrations).

- The rate constants, $r_{\ell 1}$ for Eq. (1.1-1) to r_{N3} for Eq. (1.1-6), are defined numerically, generally with the time units 1/s. These values were selected to give significant variations in all six of the PDE dependent variables, $\ell(x, t)$ to $N(x, t)$, over the time interval $0 \leq t \leq 1 \times 10^7$ s, approximately 3.85 months. Expanding this time scale is discussed subsequently corresponding to longer times for the onset of atherosclerosis.

• A spatial grid of 26 points is defined for $x_l = 0 \leq x \leq x_u = 0.004$ cm (40 μm), so that $x = 0, 0.004/25, ..., 0.004$.

```
#
# Spatial grid (in x)
  nx=26;
  xl=0;xu=0.004;
  x=seq(from=xl,to=xu,(xu-xl)/(nx-1));
```

The epithelium inner layer (EIL, intima), is thin, 0.004 cm = 40 μm.

• An interval in t of 51 points is defined for $0 \leq t \leq 1 \times 10^7$ s so that $tout = 0, 1 \times 10^7/50, ..., 1 \times 10^7$.

```
#
# Independent variable for ODE integration
  t0=0;tf=1.0e+07;nout=51;
  tout=seq(from=t0,to=tf,by=(tf-t0)/(nout-1));
```

• ICs (1.3) are defined.

```
#
# Initial condition (t=0)
  u0=rep(0,6*nx);
  for(i in 1:nx){
```

```
    u0[i]      =0;
    u0[i+nx]   =0;
    u0[i+2*nx]=0;
    u0[i+3*nx]=0;
    u0[i+4*nx]=0;
    u0[i+5*nx]=0;
  }
  ncall=0;
```

u0 therefore has $(6)(26) = 156$ elements. The counter for the calls to the ODE/MOL routine pde1a is also initialized.

- The system of 156 MOL/ODEs is integrated by the library integrator lsodes (available in deSolve, [2]). As expected, the inputs to lsodes are the ODE function, pde1a, the IC vector u0, and the vector of output values of t, tout. The length of u0 (156) informs lsodes how many ODEs are to be integrated. func,y,times are reserved names.

```
#
# ODE integration
  out=lsodes(y=u0,times=tout,func=pde1a,
      sparsetype="sparseint",rtol=1e-6,
      atol=1e-6,maxord=5);
  nrow(out)
  ncol(out)
```

The numerical solution to the ODEs is returned in matrix out. In this case, out has the dimensions $nout \times (6nx + 1) = 51 \times 6(26) + 1 = 8007$, which are confirmed by the output from nrow(out),ncol(out) (included in the numerical output considered subsequently).

The offset $156 + 1$ is required since the first element of each column has the output t (also in tout), and the $2, ..., 6nx + 1 = 2, ..., 157$ column elements have the 156 ODE solutions.

- The solutions of the 156 ODEs returned in out by lsodes are placed in arrays l,h,p,q,m,N.

```
#
# Arrays for plotting numerical solution
  l=matrix(0,nrow=nx,ncol=nout);
  h=matrix(0,nrow=nx,ncol=nout);
  p=matrix(0,nrow=nx,ncol=nout);
  q=matrix(0,nrow=nx,ncol=nout);
```

```
      m=matrix(0,nrow=nx,ncol=nout);
      N=matrix(0,nrow=nx,ncol=nout);
      for(it in 1:nout){
        for(i in 1:nx){
          l[i,it]=out[it,i+1];
          h[i,it]=out[it,i+1+nx];
          p[i,it]=out[it,i+1+2*nx];
          q[i,it]=out[it,i+1+3*nx];
          m[i,it]=out[it,i+1+4*nx];
          N[i,it]=out[it,i+1+5*nx];
        }
      }
```

Again, the offset i+1 is required since the first element of each column of out has the value of t.

- $\ell(x,t)$ to $N(x,t)$ from Eqs. (1.1) are displayed as a function of x and t with two fors.

```
#
# Display numerical solution
  iv=seq(from=1,to=nout,by=25);
  for(it in iv){
    cat(sprintf(
      "\n                      t              x"));
    cat(sprintf(
      "\n              l(x,t)        h(x,t)"));
    cat(sprintf(
      "\n              p(x,t)        q(x,t)"));
    cat(sprintf(
      "\n              m(x,t)        N(x,t)"));
    iv=seq(from=1,to=nx,by=25);
    for(i in iv){
      cat(sprintf("\n        %12.2e %12.2e",
              tout[it],x[i]));
      cat(sprintf("\n        %12.3e %12.3e",
              l[i,it],h[i,it]));
      cat(sprintf("\n        %12.3e %12.3e",
              p[i,it],q[i,it]));
      cat(sprintf("\n        %12.3e %12.3e\n",
              m[i,it],N[i,it]));
    }
```

```
   }
```

Every 25th value of x and t is displayed with by=25.

- The number of calls to pde1a is displayed at the end of the solution.

```
  #
  # Calls to ODE routine
    cat(sprintf("\n\n ncall = %5d\n\n",ncall));
```

- Graphical (plotted) output is displayed as a function of ip.

```
  #
  # Plot PDE solutions
  #
  # l,h,p,q,m,N against x, parametrically in t
  #
    if(ip==1){
  #
  # l(x,t)
    par(mfrow=c(1,1));
    matplot(x,l,type="l",xlab="x",ylab="l(x,t)",
      lty=1,main="",lwd=2,col="black");
                     .
                     .
                     .
  #
  # N(x,t)
    par(mfrow=c(1,1));
    matplot(x,N,type="l",xlab="x",ylab="N(x,t)",
      lty=1,main="",lwd=2,col="black");
    }
  #
  # l,h,p,q,m,N against t, x = xl
  #
    if(ip==2){
  #
  # l(x,t)
    par(mfrow=c(1,1));
    matplot(tout,l[1,],type="l",xlab="t",
      ylab="l(x,t)",lty=1,main="",lwd=2,
```

```
          col="black");
                          .

                          .

                          .

    #
    # N(x,t)
      par(mfrow=c(1,1));
      matplot(tout,N[1,],type="l",xlab="t",
        ylab="N(x,t)",lty=1,main="",lwd=2,
        col="black");
      }
    #
    # l,h,p,q,m,N, 3D
    #
      if(ip==3){
      t=tout
    #
    # l(x,t)
      persp(x,t,l,theta=60,phi=45,
           xlim=c(xl,xu),ylim=c(t0,tf),xlab="x",
           ylab="t",zlab="l(x,t)");
                          .

                          .

                          .


    #
    # N(x,t)
      persp(x,t,N,theta=60,phi=45,
           xlim=c(xl,xu),ylim=c(t0,tf),xlab="x",
           ylab="t",zlab="N(x,t)");
      }
```

The graphical output for ip=1,2,3 is discussed subsequently.

This completes the discussion of the main program. The ODE/MOL routine pde1a is considered next.

2.2.2   ODE/MOL ROUTINE

Listing 2.2: ODE/MOL routine pde1a for Eqs. (1.1), (1.2)

```
pde1a=function(t,u,parms){
```

```
#
# Function pde1a computes the t derivative
# vectors of l(x,t),h(x,t),p(x,t),q(x,t),
# m(x,t),N(x,t)
#
# One vector to six vectors
  l=rep(0,nx);h=rep(0,nx);
  p=rep(0,nx);q=rep(0,nx);
  m=rep(0,nx);N=rep(0,nx);
  for(i in 1:nx){
    l[i]=u[i];
    h[i]=u[i+nx];
    p[i]=u[i+2*nx];
    q[i]=u[i+3*nx];
    m[i]=u[i+4*nx];
    N[i]=u[i+5*nx];
  }
#
# lx,hx,px,qx,mx,Nx
  lx=rep(0,nx);hx=rep(0,nx);
  px=rep(0,nx);qx=rep(0,nx);
  mx=rep(0,nx);Nx=rep(0,nx);
  tablel=splinefun(x,l);lx=tablel(x,deriv=1);
  tableh=splinefun(x,h);hx=tableh(x,deriv=1);
  tablep=splinefun(x,p);px=tablep(x,deriv=1);
  tableq=splinefun(x,q);qx=tableq(x,deriv=1);
  tablem=splinefun(x,m);mx=tablem(x,deriv=1);
  tableN=splinefun(x,N);Nx=tableN(x,deriv=1);
#
# BCs
  lx[1]=-(k_l/D_l)*(l0-l[1]);lx[nx]=0;
  hx[1]=-(k_h/D_h)*(h0-h[1]);hx[nx]=0;
  px[1]=-(k_p/D_p)*(p0-p[1]);px[nx]=0;
  qx[1]=-(k_q/D_q)*(q0-q[1]);qx[nx]=0;
  mx[1]=-(k_m/D_m)*(m0-m[1]);mx[nx]=0;
  Nx[1]=-(k_N/D_N)*(N0-N[1]);Nx[nx]=0;
#
# lxx,hxx,pxx,qxx,mxx,Nxx
  lxx=rep(0,nx);hxx=rep(0,nx);
```

```
  pxx=rep(0,nx);qxx=rep(0,nx);
  mxx=rep(0,nx);Nxx=rep(0,nx);
  tablelx=splinefun(x,lx);lxx=tablelx(x,deriv=1);
  tablehx=splinefun(x,hx);hxx=tablehx(x,deriv=1);
  tablepx=splinefun(x,px);pxx=tablepx(x,deriv=1);
  tableqx=splinefun(x,qx);qxx=tableqx(x,deriv=1);
  tablemx=splinefun(x,mx);mxx=tablemx(x,deriv=1);
  tableNx=splinefun(x,Nx);Nxx=tableNx(x,deriv=1);
#
# PDEs
  lt=rep(0,nx);ht=rep(0,nx);
  pt=rep(0,nx);qt=rep(0,nx);
  mt=rep(0,nx);Nt=rep(0,nx);
  for(i in 1:nx){
#
#   Product functions
    lm=l[i]/(r_l3+l[i])*m[i];
    hN=h[i]/(r_h3+h[i])*N[i];
#
#   l_t
    lt[i]=D_l*lxx[i]-
      r_l1*lm-r_l2*l[i];
#
#   h_t
    ht[i]=D_h*hxx[i]-
      r_h1*hN-r_h2*h[i];
#
#   p_t
    pt[i]=D_p*pxx[i]+
      r_p1*lm-r_p2*p[i];
#
#   q_t
    qt[i]=D_q*qxx[i]+
      r_q1*lm-r_q2*q[i];
#
#   m_t
    mt[i]=D_m*mxx[i]-
      r_m1*(mx[i]*lx[i]+m[i]*lxx[i])-
      r_m2*lm+r_m3*hN-r_m4*m[i]+
```

```
      r_m5*l[i];
#
#    N_t
     Nt[i]=D_N*Nxx[i]+
       r_N1*lm-r_N2*hN+
       r_N3*l[i];
   }
#
# Six vectors to one vector
  ut=rep(0,6*nx);
  for(i in 1:nx){
    ut[i]      =lt[i];
    ut[i+nx]   =ht[i];
    ut[i+2*nx]=pt[i];
    ut[i+3*nx]=qt[i];
    ut[i+4*nx]=mt[i];
    ut[i+5*nx]=Nt[i];
  }
#
# Increment calls to pde1a
  ncall <<- ncall+1;
#
# Return derivative vector
  return(list(c(ut)));
  }
```

We can note the following details about Listing 2.2.

- The function is defined.

```
   pde1a=function(t,u,parms){
#
# Function pde1a computes the t derivative
# vectors of l(x,t),h(x,t),p(x,t),q(x,t),
# m(x,t),N(x,t)
```

t is the current value of t in Eqs. (1.1). u is the 156-vector of ODE/MOL dependent variables. parm is an argument to pass parameters to pde1a (unused, but required in the argument list). The arguments must be listed in the order stated to properly interface with lsodes called in the main program of Listing 2.1. The derivative vector of the LHS of Eqs. (2.1) is calculated and returned to lsodes as explained subsequently.

- u is placed in six vectors, l,h,p,q,m,N, to facilitate the programming of Eqs. (1.1).

```
#
# One vector to six vectors
  l=rep(0,nx);h=rep(0,nx);
  p=rep(0,nx);q=rep(0,nx);
  m=rep(0,nx);N=rep(0,nx);
  for(i in 1:nx){
    l[i]=u[i];
    h[i]=u[i+nx];
    p[i]=u[i+2*nx];
    q[i]=u[i+3*nx];
    m[i]=u[i+4*nx];
    N[i]=u[i+5*nx];
  }
```

- The first derivatives $\dfrac{\partial \ell}{\partial x}, \dfrac{\partial h}{\partial x}, \dfrac{\partial p}{\partial x}, \dfrac{\partial q}{\partial x}, \dfrac{\partial m}{\partial x}, \dfrac{\partial N}{\partial x}$ are computed by the spline function splinefun which is a part of the basic R system [1, 2].

```
#
# lx,hx,px,qx,mx,Nx
  lx=rep(0,nx);hx=rep(0,nx);
  px=rep(0,nx);qx=rep(0,nx);
  mx=rep(0,nx);Nx=rep(0,nx);
  tablel=splinefun(x,l);lx=tablel(x,deriv=1);
  tableh=splinefun(x,h);hx=tableh(x,deriv=1);
  tablep=splinefun(x,p);px=tablep(x,deriv=1);
  tableq=splinefun(x,q);qx=tableq(x,deriv=1);
  tablem=splinefun(x,m);mx=tablem(x,deriv=1);
  tableN=splinefun(x,N);Nx=tableN(x,deriv=1);
```

splinefun is used in two steps. The first step computes a table of spline coefficients for a numerical vector. For example, for the vector N, the table of coefficients is computed with tableN=splinefun(x,N). The second step computes the spline approximation of the derivative $\dfrac{\partial N}{\partial x}$ from the table, Nx=tableN(x,deriv=1). deriv=1 specifies a first derivative (deriv=0 specifies the vector that is differentiated and deriv=2 specifies a second derivative of the vector).

- The Robin and Neumann BCs (1.2) are implemented (the subscripts 1,nx correspond to $x = x_l, x_u$).

```
#
# BCs
  lx[1]=-(k_l/D_l)*(l0-l[1]);lx[nx]=0;
  hx[1]=-(k_h/D_h)*(h0-h[1]);hx[nx]=0;
  px[1]=-(k_p/D_p)*(p0-p[1]);px[nx]=0;
  qx[1]=-(k_q/D_q)*(q0-q[1]);qx[nx]=0;
  mx[1]=-(k_m/D_m)*(m0-m[1]);mx[nx]=0;
  Nx[1]=-(k_N/D_N)*(N0-N[1]);Nx[nx]=0;
```

- The second derivatives $\dfrac{\partial^2 \ell}{\partial x^2}, \dfrac{\partial^2 h}{\partial x^2}, \dfrac{\partial^2 p}{\partial x^2}, \dfrac{\partial^2 q}{\partial x^2}, \dfrac{\partial^2 m}{\partial x^2}, \dfrac{\partial^2 N}{\partial x^2}$ are computed by differentiating the first derivatives (successive or stagewise differentiation).

```
#
# lxx,hxx,pxx,qxx,mxx,Nxx
  lxx=rep(0,nx);hxx=rep(0,nx);
  pxx=rep(0,nx);qxx=rep(0,nx);
  mxx=rep(0,nx);Nxx=rep(0,nx);
  tablelx=splinefun(x,lx);lxx=tablelx(x,deriv=1);
  tablehx=splinefun(x,hx);hxx=tablehx(x,deriv=1);
  tablepx=splinefun(x,px);pxx=tablepx(x,deriv=1);
  tableqx=splinefun(x,qx);qxx=tableqx(x,deriv=1);
  tablemx=splinefun(x,mx);mxx=tablemx(x,deriv=1);
  tableNx=splinefun(x,Nx);Nxx=tableNx(x,deriv=1);
```

splinefun is called with deriv=1 for a first order differentiation of a first derivative.

- Equations (1.1) are programmed according to the following steps.

  – Vectors are defined for the LHS derivatives of Eqs. (1.1), $\dfrac{\partial \ell}{\partial t}, \dfrac{\partial h}{\partial t}, \dfrac{\partial p}{\partial t}, \dfrac{\partial q}{\partial t}, \dfrac{\partial m}{\partial t}, \dfrac{\partial N}{\partial t}$

```
#
# PDEs
  lt=rep(0,nx);ht=rep(0,nx);
  pt=rep(0,nx);qt=rep(0,nx);
  mt=rep(0,nx);Nt=rep(0,nx);
```

  – The products $\dfrac{\ell m}{r_{l3} + \ell}, \dfrac{h}{r_{h3} + h}$ are computed at a particular x (with a for) for use in the subsequent programming of Eqs. (1.1).

```
  for(i in 1:nx){
#
```

```
#    Product functions
     lm=l[i]/(r_l3+l[i])*m[i];
     hN=h[i]/(r_h3+h[i])*N[i];
```

- Equation (1.1-1) is programmed in the MOL format.

```
#
#    l_t
     lt[i]=D_l*lxx[i]-
       r_l1*lm-r_l2*l[i];
```

The correspondence of the programming and the mathematical statement of the PDE is an important feature of the MOL. The nonlinearity $\dfrac{\ell m}{r_{l3} + \ell}$ is easily included (lm).

- Equation (1.1-2) is programmed.

```
#
#    h_t
     ht[i]=D_h*hxx[i]-
       r_h1*hN-r_h2*h[i];
```

- Equation (1.1-3) is programmed.

```
#
#    p_t
     pt[i]=D_p*pxx[i]+
       r_p1*lm-r_p2*p[i];
```

- Equation (1.1-4) is programmed.

```
#
#    q_t
     qt[i]=D_q*qxx[i]+
       r_q1*lm-r_q2*q[i];
```

- Equation (1.1-5) is programmed.

```
#
#    m_t
     mt[i]=D_m*mxx[i]-
       r_m1*(mx[i]*lx[i]+m[i]*lxx[i])-
       r_m2*lm+r_m3*hN-r_m4*m[i]+
       r_m5*l[i];
```

The chemotaxis diffusion term $r_{m1} \left(\dfrac{\partial m}{\partial x} \dfrac{\partial \ell}{\partial x} + m \dfrac{\partial^2 \ell}{\partial x^2} \right)$ is easily included. Both nonlinear product functions are also included (with coefficients r_{m2}, r_{m3}).

      – Equation (1.1-6) is programmed.

```
#
#    N_t
     Nt[i]=D_N*Nxx[i]+
       r_N1*lm-r_N2*hN+
       r_N3*l[i];
   }
```

      The final } concludes the for in x.

• The six vectors lt,ht,pt,qt,mt,Nt are placed in a single derivative vector ut of length $6(26) = 156$ to return to lsodes.

```
#
# Six vectors to one vector
  ut=rep(0,6*nx);
  for(i in 1:nx){
    ut[i]      =lt[i];
    ut[i+nx]   =ht[i];
    ut[i+2*nx]=pt[i];
    ut[i+3*nx]=qt[i];
    ut[i+4*nx]=mt[i];
    ut[i+5*nx]=Nt[i];
  }
```

• The counter for the calls to pde1a is incremented and returned to the main program of Listing 2.1 with <<-.

```
#
# Increment calls to pde1a
  ncall <<- ncall+1;
```

• ut is returned to lsodes as a list (required by lsodes). c is the R vector utility.

```
#
# Return derivative vector
  return(list(c(ut)));
  }
```

The final } concludes pde1a.

    The output from the main program and subordinate routine of Listings 2.1 and 2.2 are considered next.

2.2.3   MODEL OUTPUT

The output from two cases is now considered. For the first case, $\ell_0 = h_0 = 0$ (rather than l0=1; h0=3; in Listing 2.1) so that there is no LDL or HDL in the bloodstream. Consequently, there are no nonzero RHS terms that would move the solutions of Eqs. (1.1) away from homogeneous (zero) ICs. We would therefore expect that the solutions for $t > 0$ would remain at the ICs. This is confirmed with the following abbreviated output (shown in Table 2.1).

We can note the following details about this output (Table 2.1).

- The dimensions of the solution matrix out are $nout \times 6nx + 1 = 51 \times 6(26) + 1 = 157$.

```
[1]  51
```

```
[1]  157
```

The offset $+1$ results from the value of t as the first element in each of the $nout = 51$ solution vectors. These same values of t are in tout,

- ICs (1.3) ($t = 0$) are verified for the homogeneous (zero) case.

- The output is for $x = 0, 0.004/25, ..., 0.004$ as programmed in Listing 2.1 (26 values of x at each value of t with every 25th value in x displayed so $x = 0, 0.004$).

- The output is for $t = 0, 1 \times 10^7/50, ..., 1 \times 10^7$ as programmed in Listing 2.1 (51 values of t with every 25th value in t displayed).

- $\ell(x, t)$ to $N(x, t)$ remain at the ICs.

- The computational effort is modest, ncall = 315, as expected since the solutions do not change with t.

The graphical output is not included here since the solutions are constant at zero.

This case may seem trivial, but it is worth executing in the sense that if any of the PDE solutions depart from the ICs, a programming error is indicated (particularly in pde1a).

For the second case, $\ell_0 = 1, h_0 = 3$ (l0=1; h0=3; in Listing 2.1). Abbreviated numerical output follows (Table 2.2).

Table 2.1: Abbreviated numerical output for $\ell_0 = h_0 = 0$ *(Continues.)*

[1] 51

[1] 157

t	x
l(x,t)	h(x,t)
p(x,t)	q(x,t)
m(x,t)	N(x,t)
0.00e+00	0.00e+00
0.000e+00	0.000e+00
0.000e+00	0.000e+00
0.000e+00	0.000e+00
0.00e+00	4.00e-03
0.000e+00	0.000e+00
0.000e+00	0.000e+00
0.000e+00	0.000e+00
t	x
l(x,t)	h(x,t)
p(x,t)	q(x,t)
m(x,t)	N(x,t)
5.00e+06	0.00e+00
0.000e+00	0.000e+00
0.000e+00	0.000e+00
0.000e+00	0.000e+00
5.00e+06	4.00e-03
0.000e+00	0.000e+00
0.000e+00	0.000e+00
0.000e+00	0.000e+00

Table 2.1: *(Continued.)* Abbreviated numerical output for $\ell_0 = h_0 = 0$

t	x
l(x,t)	h(x,t)
p(x,t)	q(x,t)
m(x,t)	N(x,t)
1.00e+07	0.00e+00
0.000e+00	0.000e+00
0.000e+00	0.000e+00
0.000e+00	0.000e+00
1.00e+07	4.00e-03
0.000e+00	0.000e+00
0.000e+00	0.000e+00
0.000e+00	0.000e+00

ncall =   315

We can note the following details about this output (Table 2.2).

- The dimensions of the solution matrix out are again $nout \times 6nx + 1 = 51 \times 6(26) + 1 = 157$.

- The homogeneous ICs (1.3) ($t = 0$) are verified.

- The output is for $x = 0, 0.004/25, ..., 0.004$ as programmed in Listing 2.1 (26 values of x at each value of t with every 25th value in x displayed so $x = 0, 0.004$).

- The output is for $t = 0, 1 \times 10^7/50, ..., 1 \times 10^7$ as programmed in Listing 2.1 (51 values of t with every 25th value in t displayed).

- $\ell(x, t)$ to $N(x, t)$ depart from the ICs in response to $\ell_0 = 1, h_0 = 3$.

- There is little variation of the solutions with x, principally because the endothelium inner layer (EIL) is thin (40 μm).

- The computational effort is greater than for the preceding case, ncall = 872, but this value for ncase is acceptable indicating that lsodes computes the solution of the 156 MOL/ODEs efficiently.

The graphical output is in Figs. 2.1-1 to 2.1-6 (for ip=3 in the main program of List-ing 2.1).

Table 2.2: Abbreviated numerical output for $\ell_0 = 1$, $h_0 = 3$ *(Continues.)*

[1] 51

[1] 157

```
              t                 x
         l(x,t)            h(x,t)
         p(x,t)            q(x,t)
         m(x,t)            N(x,t)
       0.00e+00          0.00e+00
      0.000e+00         0.000e+00
      0.000e+00         0.000e+00
      0.000e+00         0.000e+00

       0.00e+00          4.00e-03
      0.000e+00         0.000e+00
      0.000e+00         0.000e+00
      0.000e+00         0.000e+00

              t                 x
         l(x,t)            h(x,t)
         p(x,t)            q(x,t)
         m(x,t)            N(x,t)
       5.00e+06          0.00e+00
      9.782e-01         2.937e+00
      1.380e-02         2.186e-03
      1.183e-01         2.151e-01

       5.00e+06          4.00e-03
      9.782e-01         2.937e+00
      1.380e-02         2.191e-03
      1.183e-01         2.151e-01
```

Table 2.2: *(Continued.)* Abbreviated numerical output for $\ell_0 = 1$, $h_0 = 3$

t	x
l(x,t)	h(x,t)
p(x,t)	q(x,t)
m(x,t)	N(x,t)
1.00e+07	0.00e+00
9.772e-01	2.933e+00
4.319e-02	3.196e-03
1.655e-01	4.162e-01
1.00e+07	4.00e-03
9.772e-01	2.933e+00
4.319e-02	3.203e-03
1.655e-01	4.162e-01

ncall =   872

In Figs. 2.1-1 and 2.1-2, $\ell(x,t)$, $h(x,t)$ respond rapidly to $\ell_0 = 1$, $h_0 = 3$ and reach a steady state soon after $t = 0$. This rapid response is due to the effect of BCs (1.2-1) and (1.2-3).

In Figs. 2.1-3–2.1-6, $p(x,t), q(x,t), m(x,t), N(x,t)$ respond slowly and have not reached a steady state by $t_f = 1 \times 10^7$. This slow response is due to the delayed effect of the RHS terms of Eqs. (1.1-3)–(1.1-6). BCs (1.2-5), (1.2-7), (1.2-9), and (1.2-11) have little immediate effect on the solutions of Eqs. (1.1-3)–(1.1-6). Overall, two time scales are in effect as explained in Chapter 3.

In all of the six 3D plots, the solutions are invariant in x due mainly to the thin EIL.

The output for ip=1,2 is not considered here to conserve space. These two cases are left as an exercise.

To conclude this example, since two time scales were indicated by the previous output, a solution with $t_f = 1 \times 10^7$ changed to $t_f = 1 \times 10^8$ follows to study the effect of the model time scale. That is, t0=0;tf=1.0e+07;nout=51 is changed to t0=0;tf=1.0e+08;nout=51 in Listing 2.1. The numerical output is in Table 2.3.

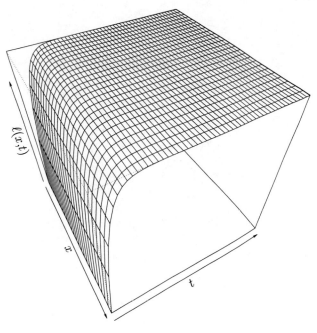

Figure 2.1-1: Numerical solution $\ell(x,t)$ from Eq. (1.1-1).

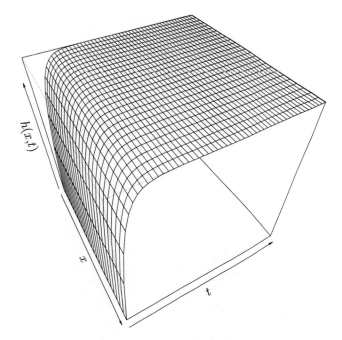

Figure 2.1-2: Numerical solution $h(x,t)$ from Eq. (1.1-2).

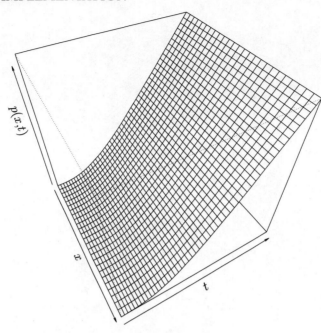

Figure 2.1-3: Numerical solution $p(x,t)$ from Eq. (1.1-3).

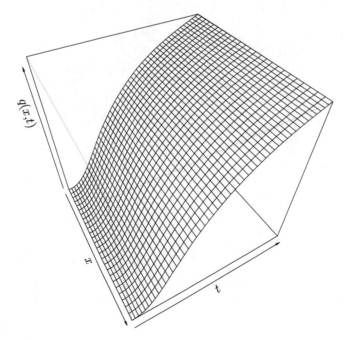

Figure 2.1-4: Numerical solution $q(x,t)$ from Eq. (1.1-4).

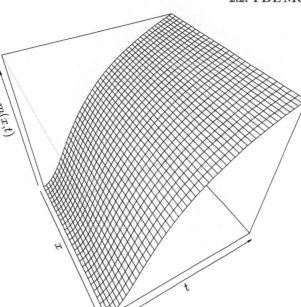

Figure 2.1-5: Numerical solution $m(x,t)$ from Eq. (1.1-5).

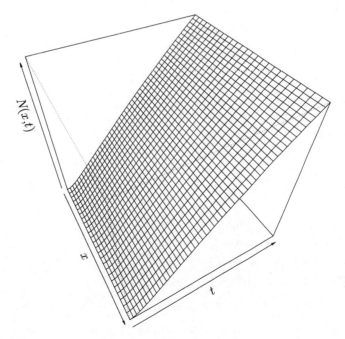

Figure 2.1-6: Numerical solution $N(x,t)$ from Eq. (1.1-6).

Table 2.3: Abbreviated numerical output for $\ell_0 = 1$, $h_0 = 3$, $t_f = 1 \times 10^8$ *(Continues.)*

[1] 51

[1] 157

t	x
l(x,t)	h(x,t)
p(x,t)	q(x,t)
m(x,t)	N(x,t)
0.00e+00	0.00e+00
0.000e+00	0.000e+00
0.000e+00	0.000e+00
0.000e+00	0.000e+00
0.00e+00	4.00e-03
0.000e+00	0.000e+00
0.000e+00	0.000e+00
0.000e+00	0.000e+00

t	x
l(x,t)	h(x,t)
p(x,t)	q(x,t)
m(x,t)	N(x,t)
5.00e+07	0.00e+00
9.754e-01	2.920e+00
2.093e-01	4.965e-03
2.535e-01	1.103e+00
5.00e+07	4.00e-03
9.754e-01	2.919e+00
2.093e-01	4.974e-03
2.535e-01	1.103e+00

Table 2.3: *(Continued.)* Abbreviated numerical output for $\ell_0 = 1$, $h_0 = 3$, $t_f = 1 \times 10^8$

t	x
l(x,t)	h(x,t)
p(x,t)	q(x,t)
m(x,t)	N(x,t)
1.00e+08	0.00e+00
9.751e-01	2.917e+00
2.611e-01	5.257e-03
2.681e-01	1.230e+00
1.00e+08	4.00e-03
9.751e-01	2.917e+00
2.611e-01	5.267e-03
2.681e-01	1.230e+00

```
ncall =   1036
```

We can note the following details about this output and the graphical output in Figs. 2.2.

- The dimensions of the solution matrix out are again $nout \times 6nx + 1 = 51 \times 6(26) + 1 = 157$.

- The homogeneous ICs (1.3) ($t = 0$) are verified.

- The output is for $x = 0, 0.004/25, ..., 0.004$ and $t = 0, 1 \times 10^8/50, ..., 1 \times 10^8$ as programmed in Listing 2.1.

- $\ell(x,t)$ and $h(x,t)$ respond essentially instantaneouly (Figs. 2.2-1 and 2.2-2) in the interval $0 \le t \le t_f = 1 \times 10^8$.

- $p(x,t)$, $q(x,t)$, $m(x,t)$, $N(x,t)$, reach a steady state (Figs. 2.2-3–2.2-6) in the interval $0 \le t \le t_f = 1 \times 10^8$.

- The computational effort is greater than for the preceding cases, ncall = 1036, reflecting the expanded time scale.

The interval $0 \le t \le t_f = 1 \times 10^8$ (approximately 3.17 years) encompasses the entire dynamic solution (from ICs (1.3) to an equilibrium). This time scale can be extended by varying the parameters defined numerically in the main program with the units of time. This variation of the model dynamic features is considered subsequently.

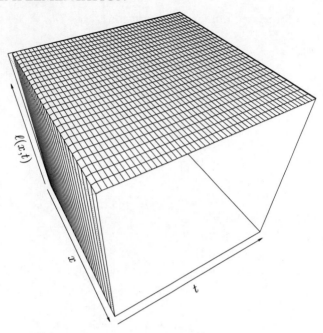

Figure 2.2-1: Numerical solution $\ell(x, t)$ from Eq. (1.1-1).

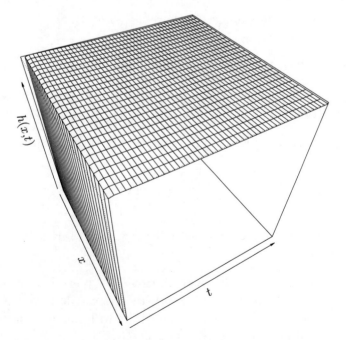

Figure 2.2-2: Numerical solution $h(x, t)$ from Eq. (1.1-2).

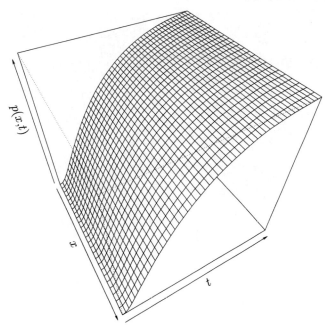

Figure 2.2-3: Numerical solution $p(x,t)$ from Eq. (1.1-3).

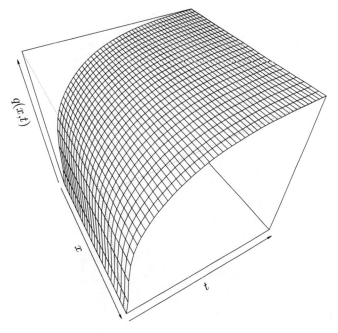

Figure 2.2-4: Numerical solution $q(x,t)$ from Eq. (1.1-4).

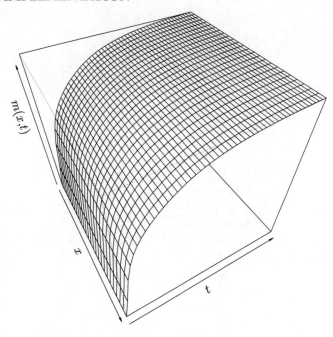

Figure 2.2-5: Numerical solution $m(x,t)$ from Eq. (1.1-5).

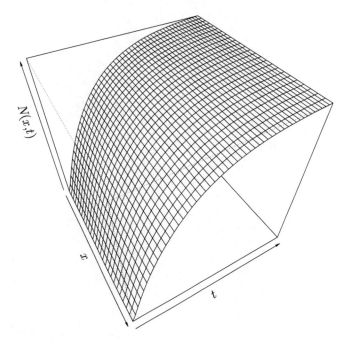

Figure 2.2-6: Numerical solution $N(x,t)$ from Eq. (1.1-6).

Since Figs. 2.1 and 2.2 do not include numerical values on the vertical (dependent variable) axis (they do not fit well in the 3D plots and therefore were not included). The values of the PDE dependent variables as a function of x and t can be observed by using ip=1,2 in the main program of Listing 2.1. This was not reported in the preceding discussion to conserve space and is left as an exercise.

2.3   SUMMARY AND CONCLUSIONS

The example in this chapter demonstrates the MOL solution of the six PDE model of Eqs. (1.1), (1.2), and (1.3) for a particular set of parameters. A remaining question is the origin of the solution features in Figs. 2.1 and 2.2. This question is addressed in Chapter 3 which pertains to a detailed analysis of the RHS terms and LHS t derivatives in Eqs. (1.1).

REFERENCES

[1] Schiesser, W. E. (2017), *Spline Collocation Methods for Partial Differential Equations: With Applications in R*, Hoboken, NJ. DOI: 10.1002/9781119301066. 28

[2] Soetaert, K., Cash, J., and Mazzia, F. (2012), *Solving Differential Equations in R*, Springer-Verlag, Heidelberg, Germany. DOI: 10.1007/978-3-642-28070-2. 21, 28

CHAPTER 3

PDE Model Detailed Analysis

3.1 INTRODUCTION

The numerical solution of Eqs. (1.1) discussed in Chapter 2 demonstrates the complicated interaction of the equations through the RHS coupling terms. For example, the programming of Eq. (1.1-5) in pde1a,

```
#
#    m_t
    mt[i]=D_m*mxx[i]-
      r_m1*(mx[i]*lx[i]+m[i]*lxx[i])-
      r_m2*lm+r_m3*hN-r_m4*m[i]+
      r_m5*l[i];
```

indicates that the LHS t derivative $\dfrac{\partial m}{\partial t}$ has the following RHS coupling:

- r_m1*(mx[i]*lx[i]+m[i]*lxx[i]) includes $\dfrac{\partial \ell}{\partial x}, \dfrac{\partial^2 \ell}{\partial x^2}$.

- r_m2*lm includes ℓ (through lm=l[i]/(r_l3+l[i])*m[i]).

- r_m3*hN includes h and N (through hN=h[i]/(r_h3+h[i])*N[i]).

- r_m5*l[i] includes ℓ.

To better understand the contributions of the various RHS terms in Eqs. (1.1), consideration is now given to how these terms can be computed and displayed as part of the numerical solution (calculation of $\ell(x,t)$, $h(x,t)$, $p(x,t)$, $q(x,t)$, $m(x,t)$, $N(x,t)$ by the method of lines (MOL) integration of Eqs. (1.1)).

The first step that is considered is the analysis of the LHS t derivatives of Eqs. (1.1). This basically is done by an extension of the main program of Listing 2.1 and a minor modification of pde1a of Listing 2.2.

3.2 PDE MODEL ROUTINES

The main program for Eqs. (1.1) (PDEs), (1.2) (BCs), and (1.3) (ICs) follows.

3.2.1   MAIN PROGRAM

Listing 3.1: Main program with analysis of the LHS t derivatives

```
#
# Six PDE atherosclerosis model
#
# Delete previous workspaces
  rm(list=ls(all=TRUE))
#
# Access ODE integrator
  library("deSolve");
#
# Access functions for numerical solution
  setwd("f:/atherosclerosis/chap3");
  source("pde1a.R");
  source("pde1b.R");
#
# Select plotting
#
#   ip=1: 2D, vs x, t parametrically
#
#   ip=2: 2D vs t, x=xl
#
#   ip=3: 3D
#
  ip=3;
#
# Parameters
#
# Diffusivities
  D_l=1.0e-08; D_h=1.0e-08;
  D_p=1.0e-08; D_q=1.0e-08;
  D_m=1.0e-08; D_N=1.0e-08;
#
# Mass transfer coefficients
  k_l=1.0e-08; k_h=1.0e-08;
  k_p=0.0e-08; k_q=1.0e-08;
  k_m=0.0e-08; k_N=0.0e-08;
#
```

```
# Bloodstream (lumen) concentrations
  l0=1; h0=3;
  p0=0; q0=0;
  m0=0; N0=0;
#
# Reaction kinetic rate constants
#
# l(x,t)
  r_l1=5.0e-08; r_l2=5.0e-08;
  r_l3=1.0e-05;
#
# h(x,t);
  r_h1=5.0e-08; r_h2=5.0e-08;
  r_h3=1.0e-05;
#
# p(x,t)
  r_p1=5.0e-08; r_p2=5.0e-08;
#
# q(x,t)
  r_q1=5.0e-08; r_q2=5.0e-08;
#
# m(x,t)
  r_m1=5.0e-08; r_m2=5.0e-08;
  r_m3=5.0e-08; r_m4=5.0e-08;
  r_m5=5.0e-08;
#
# N(x,t)
  r_N1=5.0e-08; r_N2=5.0e-08;
  r_N3=5.0e-08;
#
# Spatial grid (in x)
  nx=26;
  xl=0;xu=0.004;
  x=seq(from=xl,to=xu,(xu-xl)/(nx-1));
#
# Independent variable for ODE integration
  t0=0;tf=1.0e+07;nout=51;
  tout=seq(from=t0,to=tf,by=(tf-t0)/(nout-1));
#
```

```
# Initial condition (t=0)
  u0=rep(0,6*nx);
  for(i in 1:nx){
    u0[i]       =0;
    u0[i+nx]   =0;
    u0[i+2*nx]=0;
    u0[i+3*nx]=0;
    u0[i+4*nx]=0;
    u0[i+5*nx]=0;
  }
  ncall=0;
#
# ODE integration
  out=lsodes(y=u0,times=tout,func=pde1a,
      sparsetype="sparseint",rtol=1e-6,
      atol=1e-6,maxord=5);
  nrow(out)
  ncol(out)
#
# Arrays for plotting numerical solution
  l=matrix(0,nrow=nx,ncol=nout);
  h=matrix(0,nrow=nx,ncol=nout);
  p=matrix(0,nrow=nx,ncol=nout);
  q=matrix(0,nrow=nx,ncol=nout);
  m=matrix(0,nrow=nx,ncol=nout);
  N=matrix(0,nrow=nx,ncol=nout);
  for(it in 1:nout){
    for(i in 1:nx){
      l[i,it]=out[it,i+1];
      h[i,it]=out[it,i+1+nx];
      p[i,it]=out[it,i+1+2*nx];
      q[i,it]=out[it,i+1+3*nx];
      m[i,it]=out[it,i+1+4*nx];
      N[i,it]=out[it,i+1+5*nx];
    }
  }
#
# Display numerical solution
  iv=seq(from=1,to=nout,by=25);
```

```
   for(it in iv){
     cat(sprintf(
       "\n                     t               x"));
     cat(sprintf(
       "\n                l(x,t)          h(x,t)"));
     cat(sprintf(
       "\n                p(x,t)          q(x,t)"));
     cat(sprintf(
       "\n                m(x,t)          N(x,t)"));
   iv=seq(from=1,to=nx,by=25);
   for(i in iv){
     cat(sprintf("\n           %12.2e %12.2e",
                 tout[it],x[i]));
     cat(sprintf("\n           %12.3e %12.3e",
                 l[i,it],h[i,it]));
     cat(sprintf("\n           %12.3e %12.3e",
                 p[i,it],q[i,it]));
     cat(sprintf("\n           %12.3e %12.3e\n",
                 m[i,it],N[i,it]));
     }
   }
#
# Calls to ODE routine
   cat(sprintf("\n\n ncall = %5d\n\n",ncall));
#
# Plot PDE solutions
#
# l,h,p,q,m,N against x, parametrically in t
#
   if(ip==1){
#
# l(x,t)
   par(mfrow=c(1,1));
   matplot(x,l,type="l",xlab="x",ylab="l(x,t)",
     lty=1,main="",lwd=2,col="black");
#
# h(x,t)
   par(mfrow=c(1,1));
   matplot(x,h,type="l",xlab="x",ylab="h(x,t)",
```

```
        lty=1,main="",lwd=2,col="black");
#
# p(x,t)
  par(mfrow=c(1,1));
  matplot(x,p,type="l",xlab="x",ylab="p(x,t)",
    lty=1,main="",lwd=2,col="black");
#
# q(x,t)
  par(mfrow=c(1,1));
  matplot(x,q,type="l",xlab="x",ylab="q(x,t)",
    lty=1,main="",lwd=2,col="black");
#
# m(x,t)
  par(mfrow=c(1,1));
  matplot(x,m,type="l",xlab="x",ylab="m(x,t)",
    lty=1,main="",lwd=2,col="black");
#
# N(x,t)
  par(mfrow=c(1,1));
  matplot(x,N,type="l",xlab="x",ylab="N(x,t)",
    lty=1,main="",lwd=2,col="black");
  }
#
# l,h,p,q,m,N against t, x = xl
#
  if(ip==2){
#
# l(x,t)
  par(mfrow=c(1,1));
  matplot(tout,l[1,],type="l",xlab="t",
    ylab="l(x,t)",lty=1,main="",lwd=2,
    col="black");
#
# h(x,t)
  par(mfrow=c(1,1));
  matplot(tout,h[1,],type="l",xlab="t",
    ylab="h(x,t)",lty=1,main="",lwd=2,
    col="black");
#
```

```
# p(x,t)
  par(mfrow=c(1,1));
  matplot(tout,p[1,],type="l",xlab="t",
    ylab="p(x,t)",lty=1,main="",lwd=2,
    col="black");
#
# q(x,t)
  par(mfrow=c(1,1));
  matplot(tout,q[1,],type="l",xlab="t",
    ylab="q(x,t)",lty=1,main="",lwd=2,
    col="black");
#
# m(x,t)
  par(mfrow=c(1,1));
  matplot(tout,m[1,],type="l",xlab="t",
    ylab="m(x,t)",lty=1,main="",lwd=2,
    col="black");
#
# N(x,t)
  par(mfrow=c(1,1));
  matplot(tout,N[1,],type="l",xlab="t",
    ylab="N(x,t)",lty=1,main="",lwd=2,
    col="black");
  }
#
# l,h,p,q,m,N,  3D
#
  if(ip==3){
  t=tout
#
# l(x,t)
  persp(x,t,l,theta=60,phi=45,
       xlim=c(xl,xu),ylim=c(t0,tf),xlab="x",
       ylab="t",zlab="l(x,t)");
#
# h(x,t)
  persp(x,t,h,theta=60,phi=45,
       xlim=c(xl,xu),ylim=c(t0,tf),xlab="x",
       ylab="t",zlab="h(x,t)");
```

```
#
# p(x,t)
  persp(x,t,p,theta=60,phi=45,
        xlim=c(xl,xu),ylim=c(t0,tf),xlab="x",
        ylab="t",zlab="p(x,t)");
#
# q(x,t)
  persp(x,t,q,theta=60,phi=45,
        xlim=c(xl,xu),ylim=c(t0,tf),xlab="x",
        ylab="t",zlab="q(x,t)");
#
# m(x,t)
  persp(x,t,m,theta=60,phi=45,
        xlim=c(xl,xu),ylim=c(t0,tf),xlab="x",
        ylab="t",zlab="m(x,t)");
#
# N(x,t)
  persp(x,t,N,theta=60,phi=45,
        xlim=c(xl,xu),ylim=c(t0,tf),xlab="x",
        ylab="t",zlab="N(x,t)");
  }
#
# Plot LHS t derivaives
#
# Composite solution vector
  u=matrix(0,nrow=6*nx,ncol=nout);
  for(it in 1:nout){
  for( i in 1:nx){
    u[i,it]      =l[i,it];
    u[i+nx,it]   =h[i,it];
    u[i+2*nx,it]=p[i,it];
    u[i+3*nx,it]=q[i,it];
    u[i+4*nx,it]=m[i,it];
    u[i+5*nx,it]=N[i,it];
  }
  }
#
# Composite t derivative vector
  ut=matrix(0,nrow=6*nx,ncol=nout);
```

```
  for(it in 1:nout){
    ut[,it]=pde1b(t[it],u[,it],parms);
  }
#
# Composite t derivative vector placed
# in six vectors
  lt=matrix(0,nrow=nx,ncol=nout);
  ht=matrix(0,nrow=nx,ncol=nout);
  pt=matrix(0,nrow=nx,ncol=nout);
  qt=matrix(0,nrow=nx,ncol=nout);
  mt=matrix(0,nrow=nx,ncol=nout);
  Nt=matrix(0,nrow=nx,ncol=nout);
  for(it in 1:nout){
  for( i in 1:nx){
    lt[i,it]=ut[i,it];
    ht[i,it]=ut[i+nx,it];
    pt[i,it]=ut[i+2*nx,it];
    qt[i,it]=ut[i+3*nx,it];
    mt[i,it]=ut[i+4*nx,it];
    Nt[i,it]=ut[i+5*nx,it];
  }
  }
#
# Display LHS derivative vectors
  cat(sprintf("\n LHS t derivatives\n"));
  iv=seq(from=1,to=nout,by=25);
  for(it in iv){
  cat(sprintf(
    "\n\n         t          x      lt(x,t)"));
  cat(sprintf(
    "\n         t         x      ht(x,t)"));
  cat(sprintf(
    "\n         t         x      pt(x,t)"));
  cat(sprintf(
    "\n         t         x      qt(x,t)"));
  cat(sprintf(
    "\n         t         x      mt(x,t)"));
  cat(sprintf(
    "\n         t         x      Nt(x,t)"));
```

```
  iv=seq(from=1,to=nx,by=25);
  for(i in iv){
    cat(sprintf("\n%9.2e%11.2e%12.3e",
      t[it],x[i],lt[i,it]));
    cat(sprintf("\n%9.2e%11.2e%12.3e",
      t[it],x[i],ht[i,it]));
    cat(sprintf("\n%9.2e%11.2e%12.3e",
      t[it],x[i],pt[i,it]));
    cat(sprintf("\n%9.2e%11.2e%12.3e",
      t[it],x[i],qt[i,it]));
    cat(sprintf("\n%9.2e%11.2e%12.3e",
      t[it],x[i],mt[i,it]));
    cat(sprintf("\n%9.2e%11.2e%12.3e\n",
      t[it],x[i],Nt[i,it]));
  }
  }
#
# Plot LHS derivative vectors
#
# lt(z,t)
  persp(x,t,lt,theta=45,phi=45,
        xlim=c(xl,xu),ylim=c(t0,tf),xlab="x",
        ylab="t",zlab="lt(x,t)");
#
# ht(z,t)
  persp(x,t,ht,theta=45,phi=45,
        xlim=c(xl,xu),ylim=c(t0,tf),xlab="x",
        ylab="t",zlab="ht(x,t)");
#
# pt(z,t)
  persp(x,t,pt,theta=45,phi=45,
        xlim=c(xl,xu),ylim=c(t0,tf),xlab="x",
        ylab="t",zlab="pt(x,t)");
#
# qt(z,t)
  persp(x,t,qt,theta=45,phi=45,
        xlim=c(xl,xu),ylim=c(t0,tf),xlab="x",
        ylab="t",zlab="qt(x,t)");
#
```

```
# mt(z,t)
  persp(x,t,mt,theta=45,phi=45,
        xlim=c(xl,xu),ylim=c(t0,tf),xlab="x",
        ylab="t",zlab="mt(x,t)");
#
# Nt(z,t)
  persp(x,t,Nt,theta=45,phi=45,
        xlim=c(xl,xu),ylim=c(t0,tf),xlab="x",
        ylab="t",zlab="Nt(x,t)");
```

The main program of Listing 3.1 is just a modification of the main program of Listing 2.1 so only the differences and extensions are mentioned next.

- The accessed (external) files are specified.

```
#
# Access functions for numerical solution
  setwd("f:/atherosclerosis/chap3");
  source("pde1a.R");
  source("pde1b.R");
```

Two versions of the MOL/ODE routine are included. pde1a is the same as in Listing 2.2. pde1b is a small variation of pde1a as explained subsequently.

- An interval in t of 51 points is defined for $0 \leq t \leq 1 \times 10^7$ s so that $tout = 0, 1 \times 10^7/50, ..., 1 \times 10^7$.

```
#
# Independent variable for ODE integration
  t0=0;tf=1.0e+07;nout=51;
  tout=seq(from=t0,to=tf,by=(tf-t0)/(nout-1));
```

- pde1a is called by lsodes for the integration of the 156 MOL/ODEs (see the discussion after Listing 2.1 for details).

```
#
# ODE integration
  out=lsodes(y=u0,times=tout,func=pde1a,
        sparsetype="sparseint",rtol=1e-6,
        atol=1e-6,maxord=5);
  nrow(out)
  ncol(out)
```

- After completion of the plotting of the solutions $\ell(x, t)$ to $N(x, t)$ as explained in Chapter 2, the calculation of the t derivative vectors (LHSs of Eqs. (1.1)) is programmed, starting with a single vector u of length 6*nx = 6*26 = 156.

```
#
# Plot LHS t derivaives
#
# Composite solution vector
  u=matrix(0,nrow=6*nx,ncol=nout);
  for(it in 1:nout){
  for( i in 1:nx){
    u[i,it]       =l[i,it];
    u[i+nx,it]   =h[i,it];
    u[i+2*nx,it]=p[i,it];
    u[i+3*nx,it]=q[i,it];
    u[i+4*nx,it]=m[i,it];
    u[i+5*nx,it]=N[i,it];
  }
  }
```

- The composite t derivative vector, ut, is computed by a call to pde1b, which has the same input arguments as pde1a.

```
#
# Composite t derivative vector
  ut=matrix(0,nrow=6*nx,ncol=nout);
  for(it in 1:nout){
    ut[,it]=pde1b(t[it],u[,it],parms);
  }
```

The for steps through successive values of t (with index it). The subscript , is used for the 26 values of x.

The main difference in pde1b is the return of the derivative vector, ut, as a numerical vector

```
#
# Return derivative vector
  return(c(ut));
  }
```

rather than as a list in pde1a

```
#
# Return derivative vector
  return(list(c(ut)));
  }
```

as required by `lsodes`. `pde1b` is listed subsequently.

- Six t derivative vectors (the LHSs of Eq. (1.1)), $\dfrac{\partial \ell}{\partial t}, \dfrac{\partial h}{\partial t}, \dfrac{\partial p}{\partial t}, \dfrac{\partial q}{\partial t}, \dfrac{\partial m}{\partial t}, \dfrac{\partial N}{\partial t}$, are formed from the composite t derivative vector ut.

```
#
# Composite t derivative vector placed
# in six vectors
  lt=matrix(0,nrow=nx,ncol=nout);
  ht=matrix(0,nrow=nx,ncol=nout);
  pt=matrix(0,nrow=nx,ncol=nout);
  qt=matrix(0,nrow=nx,ncol=nout);
  mt=matrix(0,nrow=nx,ncol=nout);
  Nt=matrix(0,nrow=nx,ncol=nout);
  for(it in 1:nout){
  for( i in 1:nx){
    lt[i,it]=ut[i,it];
    ht[i,it]=ut[i+nx,it];
    pt[i,it]=ut[i+2*nx,it];
    qt[i,it]=ut[i+3*nx,it];
    mt[i,it]=ut[i+4*nx,it];
    Nt[i,it]=ut[i+5*nx,it];
  }
  }
```

- Abbreviated numerical output for the six t derivative vectors is produced in the same way as for the previous numerical output of the numerical solutions.

```
#
# Display LHS derivative vectors
  cat(sprintf("\n LHS t derivatives\n"));
  iv=seq(from=1,to=nout,by=25);
  for(it in iv){
  cat(sprintf(
    "\n\n        t          x      lt(x,t)"));
```

```
    cat(sprintf(
       "\n          t          x       ht(x,t)"));
    cat(sprintf(
       "\n          t          x       pt(x,t)"));
    cat(sprintf(
       "\n          t          x       qt(x,t)"));
    cat(sprintf(
       "\n          t          x       mt(x,t)"));
    cat(sprintf(
       "\n          t          x       Nt(x,t)"));
    iv=seq(from=1,to=nx,by=25);
    for(i in iv){
      cat(sprintf("\n%9.2e%11.2e%12.3e",
        t[it],x[i],lt[i,it]));
      cat(sprintf("\n%9.2e%11.2e%12.3e",
        t[it],x[i],ht[i,it]));
      cat(sprintf("\n%9.2e%11.2e%12.3e",
        t[it],x[i],pt[i,it]));
      cat(sprintf("\n%9.2e%11.2e%12.3e",
        t[it],x[i],qt[i,it]));
      cat(sprintf("\n%9.2e%11.2e%12.3e",
        t[it],x[i],mt[i,it]));
      cat(sprintf("\n%9.2e%11.2e%12.3e\n",
        t[it],x[i],Nt[i,it]));
    }
  }
```

Again, every 25th value in t and x is displayed.

- With the six t derivative vectors available, plotting can be completed. Here 3D plotting with persp is used, but plotting corresponding to ip = 1,2 can also be used.

-
```
#
# Plot LHS derivative vectors
#
# lt(z,t)
  persp(x,t,lt,theta=45,phi=45,
        xlim=c(xl,xu),ylim=c(t0,tf),xlab="x",
        ylab="t",zlab="lt(x,t)");
          .

          .
```

```
#
# Nt(z,t)
  persp(x,t,Nt,theta=45,phi=45,
        xlim=c(xl,xu),ylim=c(t0,tf),xlab="x",
        ylab="t",zlab="Nt(x,t)");
```

An alternative approach to computing the six *t* derivative vectors would be to place the calculation in the main program rather than using pde1b, but this would not be a good approach in this example because of the detailed calculations in pde1b that would essentially be duplicated in the main program.

3.2.2   ODE/MOL ROUTINES

pde1a is the same as in Listing 2.2 and is not repeated here. pde1b listed next is a small variation on pde1a with the return of a numerical derivative rather than a list as explained previously.

Listing 3.2: ODE/MOL routine pde1b

```
  pde1b=function(t,u,parms){
#
# Function pde1b computes the t derivative
# vectors of l(x,t),h(x,t),p(x,t),q(x,t),
# m(x,t),N(x,t)
#
# One vector to six vectors
  l=rep(0,nx);h=rep(0,nx);
  p=rep(0,nx);q=rep(0,nx);
  m=rep(0,nx);N=rep(0,nx);
  for(i in 1:nx){
    l[i]=u[i];
    h[i]=u[i+nx];
    p[i]=u[i+2*nx];
    q[i]=u[i+3*nx];
    m[i]=u[i+4*nx];
    N[i]=u[i+5*nx];
  }
#
# lx,hx,px,qx,mx,Nx
  lx=rep(0,nx);hx=rep(0,nx);
```

```
  px=rep(0,nx);qx=rep(0,nx);
  mx=rep(0,nx);Nx=rep(0,nx);
  tablel=splinefun(x,l);lx=tablel(x,deriv=1);
  tableh=splinefun(x,h);hx=tableh(x,deriv=1);
  tablep=splinefun(x,p);px=tablep(x,deriv=1);
  tableq=splinefun(x,q);qx=tableq(x,deriv=1);
  tablem=splinefun(x,m);mx=tablem(x,deriv=1);
  tableN=splinefun(x,N);Nx=tableN(x,deriv=1);
#
# BCs
  lx[1]=-(k_l/D_l)*(l0-l[1]);lx[nx]=0;
  hx[1]=-(k_h/D_h)*(h0-h[1]);hx[nx]=0;
  px[1]=-(k_p/D_p)*(p0-p[1]);px[nx]=0;
  qx[1]=-(k_q/D_q)*(q0-q[1]);qx[nx]=0;
  mx[1]=-(k_m/D_m)*(m0-m[1]);mx[nx]=0;
  Nx[1]=-(k_N/D_N)*(N0-N[1]);Nx[nx]=0;
#
# lxx,hxx,pxx,qxx,mxx,Nxx
  lxx=rep(0,nx);hxx=rep(0,nx);
  pxx=rep(0,nx);qxx=rep(0,nx);
  mxx=rep(0,nx);Nxx=rep(0,nx);
  tablelx=splinefun(x,lx);lxx=tablelx(x,deriv=1);
  tablehx=splinefun(x,hx);hxx=tablehx(x,deriv=1);
  tablepx=splinefun(x,px);pxx=tablepx(x,deriv=1);
  tableqx=splinefun(x,qx);qxx=tableqx(x,deriv=1);
  tablemx=splinefun(x,mx);mxx=tablemx(x,deriv=1);
  tableNx=splinefun(x,Nx);Nxx=tableNx(x,deriv=1);
#
# PDEs
  lt=rep(0,nx);ht=rep(0,nx);
  pt=rep(0,nx);qt=rep(0,nx);
  mt=rep(0,nx);Nt=rep(0,nx);
  for(i in 1:nx){
#
#    Product functions
    lm=l[i]/(r_l3+l[i])*m[i];
    hN=h[i]/(r_h3+h[i])*N[i];
#
#    l_t
```

```
   lt[i]=D_l*lxx[i]-
     r_l1*lm-r_l2*l[i];
#
#   h_t
   ht[i]=D_h*hxx[i]-
     r_h1*hN-r_h2*h[i];
#
#   p_t
   pt[i]=D_p*pxx[i]+
     r_p1*lm-r_p2*p[i];
#
#   q_t
   qt[i]=D_q*qxx[i]+
     r_q1*lm-r_q2*q[i];
#
#   m_t
   mt[i]=D_m*mxx[i]-
     r_m1*(mx[i]*lx[i]+m[i]*lxx[i])-
     r_m2*lm+r_m3*hN-r_m4*m[i]+
     r_m5*l[i];
#
#   N_t
   Nt[i]=D_N*Nxx[i]+
     r_N1*lm-r_N2*hN+
     r_N3*l[i];
  }
#
# Six vectors to one vector
  ut=rep(0,6*nx);
  for(i in 1:nx){
    ut[i]      =lt[i];
    ut[i+nx]   =ht[i];
    ut[i+2*nx]=pt[i];
    ut[i+3*nx]=qt[i];
    ut[i+4*nx]=mt[i];
    ut[i+5*nx]=Nt[i];
  }
#
# Return derivative vector
```

```
    return(c(ut));
    }
```

This completes the coding for Eqs. (1.1), (1.2), and (1.3), including the t derivative vectors.

3.2.3   MODEL OUTPUT

The numerical output from Listings 3.1, 2.2, and 3.2 appear in Table 3.1.

The first part of this output is a repeat of the solution in Table 2.2 (for $t_f = 1 \times 10^7$) and therefore is not included here. The second part with values of the six t derivative vectors at $t = 0, 5 \times 10^6, 1 \times 10^7$ indicates the derivatives can be positive (for PDE dependent variables increasing in t) or negative (for PDE dependent variables decreasing with t). Also, the small values of these derivatives for $t > 0$ is consistent with $t_f = 1 \times 10^7$.

Since the solutions for the six dependent variables are the same as in Chapter 2, Table 2.2 and Figs. (2.1), the 3D plots are not repeated here. The six t derivatives vectors are plotted in Figs. (3.1).

The essentially instantaneous response in Figs. 3.1-1 and 3.1-2 reflects the change from

(1) IC (1.3-1), $\ell(x, t = 0) = 0$, with derivative $\dfrac{\partial \ell(x, t = 0)}{\partial t} = 1.164 \times 10^{-4}$ (Table 3.1), and

(2) IC (1.3-2), $h(x, t = 0) = 0$, with derivative $\dfrac{\partial h(x, t = 0)}{\partial t} = 3.491 \times 10^{-4}$ (Table 3.1), to smaller values of the derivatives for $t > 0$.

In summary, as indicated in Figs. (2.2), not all of the six PDE dependent variables have reached a steady state with $t_f = 1 \times 10^7$. Why this occurs can be addressed by examining the RHS PDE terms in detail. This is illustrated with a following example for Eq. (1.1-6) since foam cells (with density $N(x, t)$) play a central role in the pathology of atherosclerosis.

Table 3.1: Numerical output from the routines in Listings 3.1, 2.2, 3.2 *(Continues.)*

[1] 51

[1] 157

t	x	lt(x,t)
t	x	ht(x,t)
t	x	pt(x,t)
t	x	qt(x,t)
t	x	mt(x,t)
t	x	Nt(x,t)
0.00e+00	0.00e+00	1.164e-04
0.00e+00	0.00e+00	3.491e-04
0.00e+00	0.00e+00	0.000e+00
0.00e+00	0.00e+00	0.000e+00
0.00e+00	0.00e+00	0.000e+00
0.00e+00	0.00e+00	0.000e+00
0.00e+00	4.00e-03	1.692e-18
0.00e+00	4.00e-03	5.077e-18
0.00e+00	4.00e-03	0.000e+00
0.00e+00	4.00e-03	0.000e+00
0.00e+00	4.00e-03	0.000e+00
0.00e+00	4.00e-03	0.000e+00

Table 3.1: *(Continued.)* Numerical output from the routines in Listings 3.1, 2.2, 3.2

```
          t            x        lt(x,t)
          t            x        ht(x,t)
          t            x        pt(x,t)
          t            x        qt(x,t)
          t            x        mt(x,t)
          t            x        Nt(x,t)
    5.00e+06     0.00e+00    -3.352e-10
    5.00e+06     0.00e+00    -8.547e-10
    5.00e+06     0.00e+00     5.228e-09
    5.00e+06     0.00e+00     3.419e-10
    5.00e+06     0.00e+00     1.559e-08
    5.00e+06     0.00e+00     4.407e-08

    5.00e+06     4.00e-03    -3.359e-10
    5.00e+06     4.00e-03    -8.564e-10
    5.00e+06     4.00e-03     5.228e-09
    5.00e+06     4.00e-03     3.425e-10
    5.00e+06     4.00e-03     1.559e-08
    5.00e+06     4.00e-03     4.407e-08
```

Table 3.1: *(Continued.)* Numerical output from the routines in Listings 3.1, 2.2, 3.2

t	x	lt(x,t)
t	x	ht(x,t)
t	x	pt(x,t)
t	x	qt(x,t)
t	x	mt(x,t)
t	x	Nt(x,t)
1.00e+07	0.00e+00	-1.231e-10
1.00e+07	0.00e+00	-7.240e-10
1.00e+07	0.00e+00	6.115e-09
1.00e+07	0.00e+00	1.231e-10
1.00e+07	0.00e+00	5.949e-09
1.00e+07	0.00e+00	3.632e-08
1.00e+07	4.00e-03	-1.234e-10
1.00e+07	4.00e-03	-7.255e-10
1.00e+07	4.00e-03	6.115e-09
1.00e+07	4.00e-03	1.234e-10
1.00e+07	4.00e-03	5.949e-09
1.00e+07	4.00e-03	3.632e-08

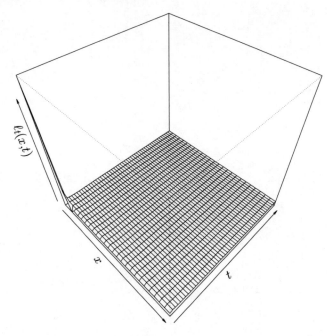

Figure 3.1-1: Numerical solution $\ell_t(x,t)$ from Eq. (1.1-1), $t_f = 1 \times 10^7$.

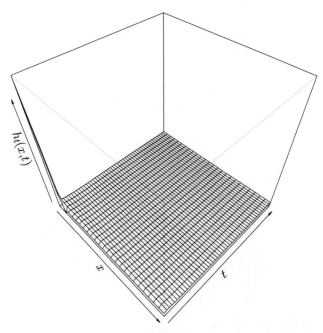

Figure 3.1-2: Numerical solution $h_t(x,t)$ from Eq. (1.1-2), $t_f = 1 \times 10^7$.

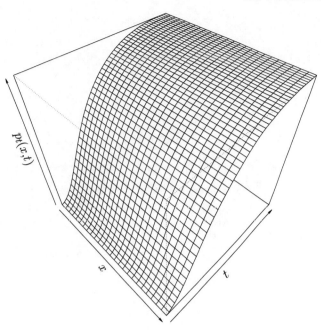

Figure 3.1-3: Numerical solution $p_t(x, t)$ from Eq. (1.1-3), $t_f = 1 \times 10^7$.

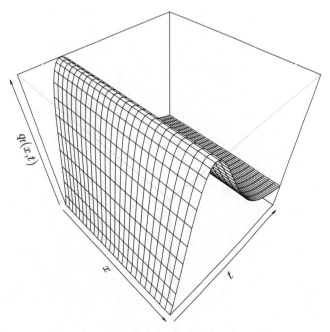

Figure 3.1-4: Numerical solution $q_t(x, t)$ from Eq. (1.1-4), $t_f = 1 \times 10^7$.

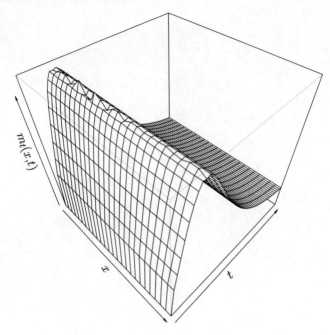

Figure 3.1-5: Numerical solution $m_t(x, t)$ from Eq. (1.1-5), $t_f = 1 \times 10^7$.

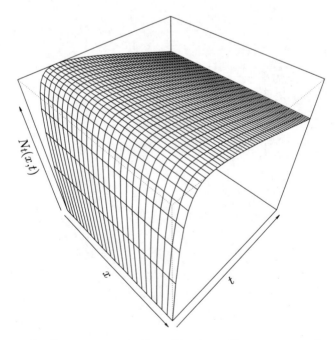

Figure 3.1-6: Numerical solution $N_t(x, t)$ from Eq. (1.1-6), $t_f = 1 \times 10^7$.

To complete this discussion, the t interval is extended with

```
t0=0;tf=1.0e+07;nout=51;
```

changed in Listing 3.1 to

```
t0=0;tf=1.0e+08;nout=51;
```

The numerical output is not included here to conserve space, other than to point out the change from `ncall = 872` to `ncall = 1036` as in Tables 2.2 and 2.3. This modest increase results from the extended t interval.

The 3D graphical output is in Figs. (3.2).

We can note the following details about this output:

- All of the derivatives in t approach closer to a steady state than in Figs. (3.1).

- $q_t(x,t)$ in Fig. 3.2-4 has a temporary nonsmooth condition for a narrow interval in t that is then smoothed for subsequent t. This irregularity in $q_t(x,t)$ does not appreciably affect $q(x,t)$ as reflected in Fig. 2.2-4 since the integration in t (from $q_t(x,t)$ to $q(x,t)$) is a smoothing process.

- Irregularities in t derivatives (and also, x derivatives as demonstrated in the next example) are not uncommon and do not necessarily carry over to the integrated variables. The origin and effect of irregular derivatives can be studied in detail as illustrated by the following example.

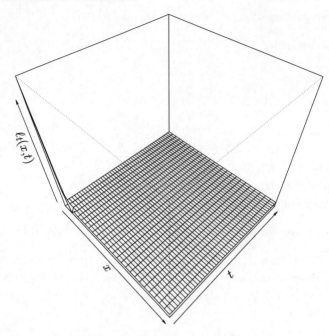

Figure 3.2-1: Numerical solution $\ell_t(x, t)$ from Eq. (1.1-1), $t_f = 1 \times 10^8$.

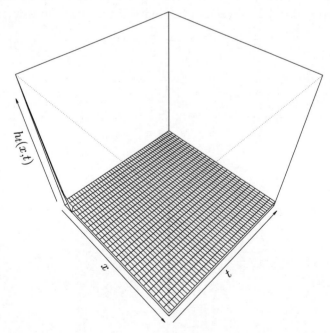

Figure 3.2-2: Numerical solution $h_t(x, t)$ from Eq. (1.1-2), $t_f = 1 \times 10^8$.

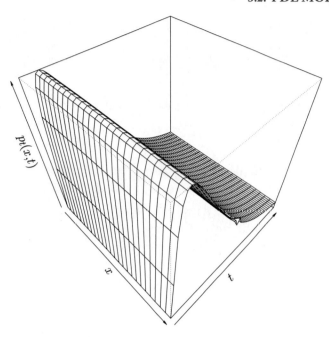

Figure 3.2-3: Numerical solution $p_t(x, t)$ from Eq. (1.1-3), $t_f = 1 \times 10^8$.

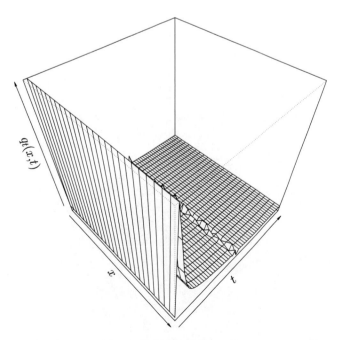

Figure 3.2-4: Numerical solution $q_t(x, t)$ from Eq. (1.1-4), $t_f = 1 \times 10^8$.

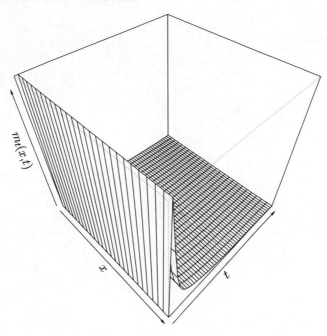

Figure 3.2-5: Numerical solution $m_t(x, t)$ from Eq. (1.1-5), $t_f = 1 \times 10^8$.

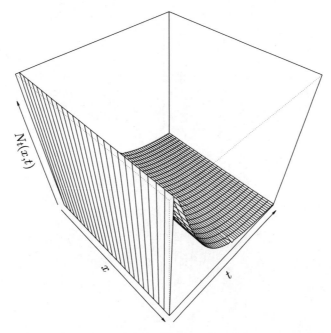

Figure 3.2-6: Numerical solution $N_t(x, t)$ from Eq. (1.1-6), $t_f = 1 \times 10^8$.

3.2.4 MAIN PROGRAM WITH DETAILED PDE ANALYSIS

To study the RHS terms of Eq. (1.1-6), the following code is added to the end of the main program of Listing 3.1.

Listing 3.3: Addition to the main program of Listing 3.1

```
          .
          .
          .

Main program of Listing 3.1

          .
          .
          .

#
# Compute RHS terms, N(x,t)
      Nx=matrix(0,nrow=nx,ncol=nout);
     Nxx=matrix(0,nrow=nx,ncol=nout);
  cat(sprintf("\n RHS terms, N(x,t)\n"));
  for(it in 1:nout){
#
# Nx
  tableN=splinefun(x,N[,it]);
  Nx[,it]=tableN(x,deriv=1);
#
# BCs, x=xl,xu
  Nx[ 1,it]=-(k_N/D_N)*(N0-N[1,it]);
  Nx[nx,it]=0;
#
# Nxx
  tableNx=splinefun(x,Nx[,it]);
  Nxx[,it]=tableNx(x,deriv=1);
  }
#
# D_N*Nxx
  term1=matrix(0,nrow=nx,ncol=nout);
  for(it in 1:nout){
  for( i in 1:nx)  {
    term1[i,it]=D_N*Nxx[i,it];
  }
  }
```

```
#
# r_N 1*(l/(r_l3+l))*m
  term2=matrix(0,nrow=nx,ncol=nout);
  for(it in 1:nout){
  for( i in 1:nx)  {
    term2[i,it]=r_N1*(l[i,it]/(r_l3+l[i,it]))*
                       m[i,it];
  }
  }
#
# -r_N 2*(h/(r_h3+h))*N
  term3=matrix(0,nrow=nx,ncol=nout);
  for(it in 1:nout){
  for( i in 1:nx)  {
    term3[i,it]=-r_N2*(h[i,it]/(r_h3+h[i,it]))*
                       N[i,it];
  }
  }
#
# r_N3*l
  term4=matrix(0,nrow=nx,ncol=nout);
  for(it in 1:nout){
  for( i in 1:nx)  {
    term4[i,it]=r_N3*l[i,it];
  }
  }
#
# Nt
  term5=matrix(0,nrow=nx,ncol=nout);
  for(it in 1:nout){
  for( i in 1:nx)  {
    term5[i,it]=term1[i,it]+term2[i,it]+
                term3[i,it]+term4[i,it];
  }
  }
#
# Display terms in N(x,t) PDE
  iv=seq(from=1,to=nout,by=25);
  for(it in iv){
```

```
  cat(sprintf(
    "\n\n          t          x          term1"));
  cat(sprintf(
    "\n          t          x          term2"));
  cat(sprintf(
    "\n          t          x          term3"));
  cat(sprintf(
    "\n          t          x          term4"));
  cat(sprintf(
    "\n          t          x          term5"));
  iv=seq(from=1,to=nx,by=5);
  for(i in iv){
    cat(sprintf("\n%9.2e%11.2e%12.3e",
      t[it],x[i],term1[i,it]));
    cat(sprintf("\n%9.2e%11.2e%12.3e",
      t[it],x[i],term2[i,it]));
    cat(sprintf("\n%9.2e%11.2e%12.3e",
      t[it],x[i],term3[i,it]));
    cat(sprintf("\n%9.2e%11.2e%12.3e",
      t[it],x[i],term4[i,it]));
    cat(sprintf("\n%9.2e%11.2e%12.3e\n",
      t[it],x[i],term5[i,it]));
  }
  }
#
# Plot RHS terms, N(x,t)
#
# term1
  persp(x,t,term1,theta=45,phi=45,
        xlim=c(xl,xu),ylim=c(t0,tf),xlab="x",
        ylab="t",zlab="term1");
#
# term2
  persp(x,t,term2,theta=45,phi=45,
        xlim=c(xl,xu),ylim=c(t0,tf),xlab="x",
        ylab="t",zlab="term2");
#
# term3
  persp(x,t,term3,theta=45,phi=45,
```

```
              xlim=c(xl,xu),ylim=c(t0,tf),xlab="x",
              ylab="t",zlab="term3");
#
# term4
   persp(x,t,term4,theta=45,phi=45,
          xlim=c(xl,xu),ylim=c(t0,tf),xlab="x",
          ylab="t",zlab="term4");
#
# Plot LHS term, N(x,t)
   persp(x,t,term5,theta=45,phi=45,
          xlim=c(xl,xu),ylim=c(t0,tf),xlab="x",
          ylab="t",zlab="term5");
```

We can note the following details about this code.

- Matrices Nx,Nxx are defined for $\dfrac{\partial N}{\partial x}, \dfrac{\partial^2 N}{\partial x^2}$.

```
   #
   # Compute RHS terms, N(x,t)
        Nx=matrix(0,nrow=nx,ncol=nout);
        Nxx=matrix(0,nrow=nx,ncol=nout);
     cat(sprintf("\n RHS terms, N(x,t)\n"));
```

- splinefun is used to calculate $\dfrac{\partial N}{\partial x}, \dfrac{\partial^2 N}{\partial x^2}$, including BCs (1.2-11) and (1.2-12).

```
     for(it in 1:nout){
   #
   # Nx
     tableN=splinefun(x,N[,it]);
     Nx[,it]=tableN(x,deriv=1);
   #
   # BCs, x=xl,xu
     Nx[ 1,it]=-(k_N/D_N)*(N0-N[1,it]);
     Nx[nx,it]=0;
   #
   # Nxx
     tableNx=splinefun(x,Nx[,it]);
     Nxx[,it]=tableNx(x,deriv=1);
     }
```

Since k_N=0 (in Listing 3.1), there is no transfer of the foam cells from the EIL to the bloodstream.

- $D_N \dfrac{\partial^2 N(x,t)}{\partial x^2}$ in Eq. (1.1-6) is computed as `term1` and plotted below in 3D in Fig. 3.3-1.

```
#
# D_N*Nxx
  term1=matrix(0,nrow=nx,ncol=nout);
  for(it in 1:nout){
  for( i in 1:nx)  {
    term1[i,it]=D_N*Nxx[i,it];
  }
  }
```

`term1` represents the diffusion of foam cells within the epithelium inner layer (EIL) and is small since $N(x,t)$ is nearly uniform across the EIL (so that `Nxx` is small).

- $r_{N1} \dfrac{\ell(x,t)}{r_{l3} + \ell(x,t)} m(x,t)$ in Eq. (1.1-6) is computed as `term2` and plotted below in 3D in Fig. 3.3-2.

```
#
# r_N1*(1/(r_13+1))*m
  term2=matrix(0,nrow=nx,ncol=nout);
  for(it in 1:nout){
  for( i in 1:nx)  {
    term2[i,it]=r_N1*(l[i,it]/(r_13+l[i,it]))*
                     m[i,it];
  }
  }
```

`term2` represents the production of foam cells ($N(x,t)$) by the simultaneous action of modified LDL ($\ell(x,t)$) and macrophages ($m(x,t)$). r_{l3} has a relatively large value (`1.0e-05`) to avoid a division by a small number.

- $-r_{N2} \dfrac{h(x,t)}{r_{h3} + h(x,t)} N(x,t)$ in Eq. (1.1-6) is computed as `term3` and plotted below in 3D in Fig. 3.3-3.

```
#
# -r_N2*(h/(r_h3+h))*N
  term3=matrix(0,nrow=nx,ncol=nout);
```

```
      for(it in 1:nout){
      for( i in 1:nx)   {
        term3[i,it]=-r_N2*(h[i,it]/(r_h3+h[i,it]))*
                          N[i,it];
      }
      }
```

`term3` represents the removal of foam cells by the simultaneous action of HDL ($h(x,t)$) and foam cells ($N(x,t)$). r_{h3} has a relatively large value (`1.0e-05`) to avoid a division by a small number.

- $r_{N3}\ell(x,t)$ in Eq. (1.1-6) is computed as `term4` and plotted below in 3D in Fig. 3.3-4.

```
      #
      # r_N3*l
        term4=matrix(0,nrow=nx,ncol=nout);
        for(it in 1:nout){
        for( i in 1:nx)   {
          term4[i,it]=r_N3*l[i,it];
        }
        }
```

`term4` represents the production of foam cells directly by modified LDL.

- $\dfrac{\partial N(x,t)}{\partial t}$, the LHS of Eq. (1.1-6), is computed as `term5` by summing the RHS terms `term1` to `term4` and plotted below in 3D in Fig. 3.3-5.

```
      #
      # Nt
        term5=matrix(0,nrow=nx,ncol=nout);
        for(it in 1:nout){
        for( i in 1:nx)   {
          term5[i,it]=term1[i,it]+term2[i,it]+
                      term3[i,it]+term4[i,it];
        }
        }
```

- Selected values of `term1` to `term5` are displayed (with by=25 for t and by=5 for x).

```
      #
      # Display terms in N(x,t) PDE
```

```
  iv=seq(from=1,to=nout,by=25);
  for(it in iv){
  cat(sprintf(
    "\n\n        t           x           term1"));
  cat(sprintf(
    "\n         t           x           term2"));
  cat(sprintf(
    "\n         t           x           term3"));
  cat(sprintf(
    "\n         t           x           term4"));
  cat(sprintf(
    "\n         t           x           term5"));
  iv=seq(from=1,to=nx,by=5);
  for(i in iv){
    cat(sprintf("\n%9.2e%11.2e%12.3e",
      t[it],x[i],term1[i,it]));
    cat(sprintf("\n%9.2e%11.2e%12.3e",
      t[it],x[i],term2[i,it]));
    cat(sprintf("\n%9.2e%11.2e%12.3e",
      t[it],x[i],term3[i,it]));
    cat(sprintf("\n%9.2e%11.2e%12.3e",
      t[it],x[i],term4[i,it]));
    cat(sprintf("\n%9.2e%11.2e%12.3e\n",
      t[it],x[i],term5[i,it]));
  }
  }
```

- term1 to term5 are plotted in 3D with persp.

```
  #
  # Plot RHS terms, N(x,t)
  #
  # term1
    persp(x,t,term1,theta=45,phi=45,
          xlim=c(xl,xu),ylim=c(t0,tf),xlab="x",
          ylab="t",zlab="term1");
  #
  # term2
    persp(x,t,term2,theta=45,phi=45,
          xlim=c(xl,xu),ylim=c(t0,tf),xlab="x",
```

```
            ylab="t",zlab="term2");
    #
    # term3
      persp(x,t,term3,theta=45,phi=45,
            xlim=c(xl,xu),ylim=c(t0,tf),xlab="x",
            ylab="t",zlab="term3");
    #
    # term4
      persp(x,t,term4,theta=45,phi=45,
            xlim=c(xl,xu),ylim=c(t0,tf),xlab="x",
            ylab="t",zlab="term4");
    #
    # Plot LHS term, N(x,t)
      persp(x,t,term5,theta=45,phi=45,
            xlim=c(xl,xu),ylim=c(t0,tf),xlab="x",
            ylab="t",zlab="term5");
```

This completes the coding of the RHS and LHS terms of Eq. (1.1-6). The physical interpretation of these terms, as well as of Eqs. (1.1-1) to (1.1-5), is explained in detail in [1–4].

3.2.5   ODE/MOL ROUTINES

The ODE/MOL routines are pde1a in Listing 2.2 and pde1b in Listing 3.2 (called by the main program of Listing 3.1).

3.2.6   MODEL OUTPUT

The first part of the model output is the same as in Table 2.2. The second part is the same as in Table 3.1. The third part for term1 to term5 is in Table 3.2.

The 3D plotting of term1 to term5 is in Figs. (3.3), which indicate that Eq. (1.1-6) appears to be reaching a steady state (equilibrium) solution (for $t_f = 1 \times 10^7$). Figures 3.1-6 and 3.3-5 are the same (for $\dfrac{\partial N(x,t)}{\partial t}$).

As a concluding example, the t interval is extended by

```
t0=0;tf=1.0e+07;nout=51;
```

changed in Listing 3.1 to

```
t0=0;tf=1.0e+08;nout=51;
```

Execution of the routines in Listings 3.1, 2.2, 3.2, and 3.3 gives the following abbreviated output for the third part of the numerical output (plotting of term1 to term5); see Table 3.3.

Table 3.2: Abbreviated numerical output from Listing 3.3 *(Continues.)*

t	x	term1
t	x	term2
t	x	term3
t	x	term4
t	x	term5
0.00e+00	0.00e+00	0.000e+00
0.00e+00	0.00e+00	0.000e+00
0.00e+00	0.00e+00	-0.000e+00
0.00e+00	0.00e+00	0.000e+00
0.00e+00	0.00e+00	0.000e+00

```
                  .

                  .

                  .
   Output for x=8.00e-04
    to 3.20e-03 removed

                  .

                  .

                  .
```

0.00e+00	4.00e-03	0.000e+00
0.00e+00	4.00e-03	0.000e+00
0.00e+00	4.00e-03	-0.000e+00
0.00e+00	4.00e-03	0.000e+00
0.00e+00	4.00e-03	0.000e+00

Table 3.2: *(Continued.)* Abbreviated numerical output from Listing 3.3

```
        t            x           term1
        t            x           term2
        t            x           term3
        t            x           term4
        t            x           term5
   5.00e+06    0.00e+00   -1.453e-12
   5.00e+06    0.00e+00    5.917e-09
   5.00e+06    0.00e+00   -1.076e-08
   5.00e+06    0.00e+00    4.891e-08
   5.00e+06    0.00e+00    4.407e-08
                    .
                    .
                    .

   Output for x=8.00e-04
     to 3.20e-03 removed
                    .
                    .
                    .
   5.00e+06    4.00e-03    7.265e-13
   5.00e+06    4.00e-03    5.917e-09
   5.00e+06    4.00e-03   -1.076e-08
   5.00e+06    4.00e-03    4.891e-08
   5.00e+06    4.00e-03    4.407e-08
```

Table 3.2: *(Continued.)* Abbreviated numerical output from Listing 3.3

```
       t            x          term1
       t            x          term2
       t            x          term3
       t            x          term4
       t            x          term5
  1.00e+07    0.00e+00   -1.520e-12
  1.00e+07    0.00e+00    8.274e-09
  1.00e+07    0.00e+00   -2.081e-08
  1.00e+07    0.00e+00    4.886e-08
  1.00e+07    0.00e+00    3.632e-08

              .

              .

              .

   Output for x=8.00e-04
    to 3.20e-03 removed

              .

              .

              .

  1.00e+07    4.00e-03    7.602e-13
  1.00e+07    4.00e-03    8.274e-09
  1.00e+07    4.00e-03   -2.081e-08
  1.00e+07    4.00e-03    4.886e-08
  1.00e+07    4.00e-03    3.632e-08
```

Table 3.3: Abbreviated numerical output from Listing 3.3 *(Continues.)*

```
        t           x          term1
        t           x          term2
        t           x          term3
        t           x          term4
        t           x          term5
  0.00e+00    0.00e+00     0.000e+00
  0.00e+00    0.00e+00     0.000e+00
  0.00e+00    0.00e+00    -0.000e+00
  0.00e+00    0.00e+00     0.000e+00
  0.00e+00    0.00e+00     0.000e+00

               .
               .
               .

  Output for x=8.00e-04
    to 3.20e-03 removed

               .
               .
               .

  0.00e+00    4.00e-03     0.000e+00
  0.00e+00    4.00e-03     0.000e+00
  0.00e+00    4.00e-03    -0.000e+00
  0.00e+00    4.00e-03     0.000e+00
  0.00e+00    4.00e-03     0.000e+00
```

Table 3.3: *(Continued.)* Abbreviated numerical output from Listing 3.3

```
         t            x         term1
         t            x         term2
         t            x         term3
         t            x         term4
         t            x         term5
  5.00e+07    0.00e+00    1.937e-10
  5.00e+07    0.00e+00    1.267e-08
  5.00e+07    0.00e+00   -5.517e-08
  5.00e+07    0.00e+00    4.877e-08
  5.00e+07    0.00e+00    6.466e-09

               .

               .

               .

   Output for x=8.00e-04
    to 3.20e-03 removed

               .

               .

               .

  5.00e+07    4.00e-03    6.182e-10
  5.00e+07    4.00e-03    1.267e-08
  5.00e+07    4.00e-03   -5.517e-08
  5.00e+07    4.00e-03    4.877e-08
  5.00e+07    4.00e-03    6.888e-09
```

Table 3.3: *(Continued.)* Abbreviated numerical output from Listing 3.3

```
        t            x          term1
        t            x          term2
        t            x          term3
        t            x          term4
        t            x          term5
 1.00e+08    0.00e+00   -1.633e-12
 1.00e+08    0.00e+00    1.341e-08
 1.00e+08    0.00e+00   -6.148e-08
 1.00e+08    0.00e+00    4.876e-08
 1.00e+08    0.00e+00    6.837e-10

                 .

                 .

                 .

   Output for x=8.00e-04
     to 3.20e-03 removed

                 .

                 .

                 .

 1.00e+08    4.00e-03    9.039e-13
 1.00e+08    4.00e-03    1.341e-08
 1.00e+08    4.00e-03   -6.148e-08
 1.00e+08    4.00e-03    4.875e-08
 1.00e+08    4.00e-03    6.838e-10
```

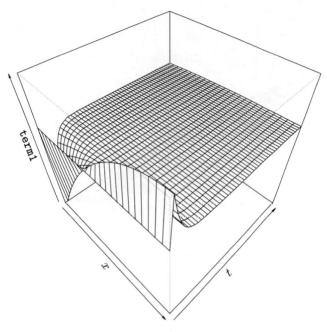

Figure 3.3-1: term1 in Eq. (1.1-6), $t_f = 1 \times 10^7$.

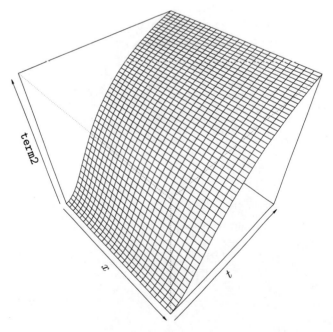

Figure 3.3-2: term2 in Eq. (1.1-6), $t_f = 1 \times 10^7$.

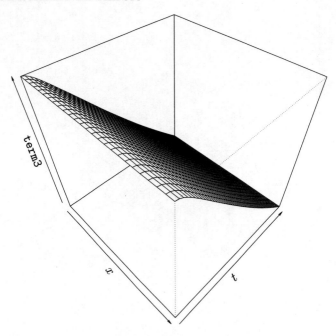

Figure 3.3-3: `term3` in Eq. (1.1-6), $t_f = 1 \times 10^7$.

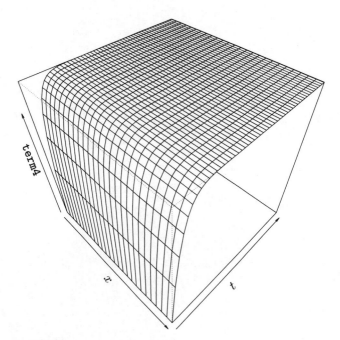

Figure 3.3-4: `term4` in Eq. (1.1-6), $t_f = 1 \times 10^7$.

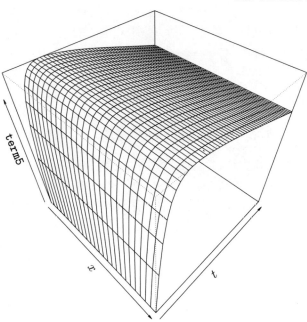

Figure 3.3-5: `term5` in Eq. (1.1-6), $t_f = 1 \times 10^7$.

The 3D plotting of `term1` to `term5` is in Figs. 3.4.

Figures (3.4) indicate that Eq. (1.1-6) has reached a steady state (equilibrium) solution (for $t_f = 1 \times 10^8$). Figures 3.2-6 and 3.4-5 are the same (for $\dfrac{\partial N(x,t)}{\partial t}$).

Figure 3.4-1 (`term1`) indicates an irregularity for a narrow interval in t that is smoothed for the remainder of the solution. This irregularity is actually small as expected (since $N(x,t)$ is nearly uniform with respect to x across the EIL so that `Nxx` is small) and has a negligible effect on $N(x,t)$ (Fig. 2.2-6). But the irregularity appears to be large from the automatic vertical z scaling in Fig. 3.4-1. For example, it can be eliminated by adding `zlim=c(-1.0e-08,1.0e+08)` to the call to `persp` for plotting `term1` (left as an excercise).

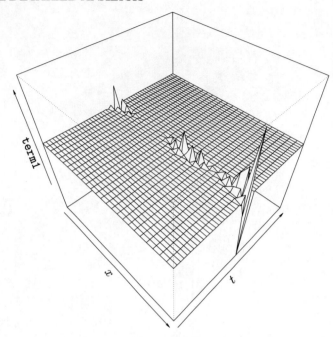

Figure 3.4-1: `term1` in Eq. (1.1-6), $t_f = 1 \times 10^8$.

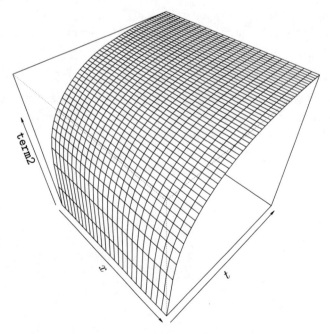

Figure 3.4-2: `term2` in Eq. (1.1-6), $t_f = 1 \times 10^8$.

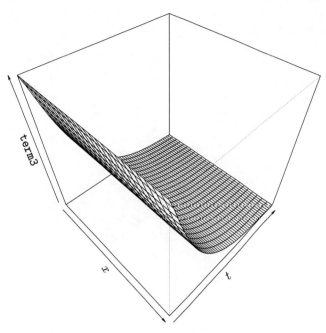

Figure 3.4-3: term3 in Eq. (1.1-6), $t_f = 1 \times 10^8$.

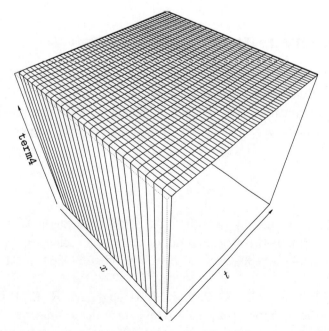

Figure 3.4-4: term4 in Eq. (1.1-6), $t_f = 1 \times 10^8$.

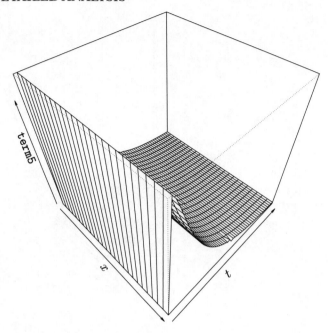

Figure 3.4-5: `term5` in Eq. (1.1-6), $t_f = 1 \times 10^8$.

3.3   SUMMARY AND CONCLUSIONS

The examples in this chapter explain a methodology for analyzing the RHS and LHS terms in a system of PDEs. This detailed analysis provides a quantitative understanding of the contributions of the various terms that serves as the basis for further model development and testing.

The next concluding chapter illustrates how computer-based experimentation can be used for model development and refinement.

REFERENCES

[1] Chalmers, A. D., Cohen, A., Bursill, C. A., and Myerscough, M. R. (2015), Bifurcation and dynamics in a mathematical model of early atherosclerosis, *Journal of Mathematical Biology*, 71, pp. 1451–1480. DOI: 10.1007/s00285-015-0864-5. 82

[2] Chalmers, A. D., Bursill, C. A., and Myerscough, M. R. (2017), Nonlinear dynamics of early atherosclerosis plaque formation may determine the effect of high density lipoproteins in plaque regressions, *Plos One*, 12, no. 11, November. DOI: 10.1371/journal.pone.0187674.

[3] Hao, W. and Friedman, A. (2014), The LDL-HDL profile determines the risk of atherosclerosis: A mathematical model, *Plos One*, 9, no. 3, March. DOI: 10.1371/journal.pone.0090497.

[4] Khatib, N. E., Genieys, S., Kazmierczak, B., and Volpert, V. (2012), Reaction-diffusion model of atherosclerosis development, *Journal of Mathematical Biology*, 65, pp. 349–374. DOI: 10.1007/s00285-011-0461-1. 82

CHAPTER 4

PDE Model Applications

4.1   INTRODUCTION

In this concluding chapter, consideration is given to the application of the PDE model developed in the earlier chapters to the therapeutic treatment of atherosclerosis through the use of LDL-lowering and HDL-raising drugs [1].

The starting point is the extension of the PDE model Eqs. (1.1), (1.2), and (1.3). Three cases are programmed as variants of BCs (1.2-1) and (1.2-3).

- ncase=1:

$$\ell_0 = 2; \; h_0 = 2$$

so that the bloodstream has equal concentrations of LDL and HDL which is considered an unhealthy condition.

- ncase=2:

$$\ell_0 = 2 - (1 - \exp(-5t/t_f)); \; h_0 = 2.$$

At $t = 0$, the LDL bloodstream concentration is $\ell_0 = 2$. This concentration decreases with increasing t, possibly in response to a LDL-lowering drug, and eventually at $t = t_f = 1 \times 10^8$, $\ell_0 = 2 - (1 - \exp(-5)) = 1.007$ so that the bloodstream concentration is lowered from 2 to 1 (while the HDL concentration remains at 2).

- ncase=3:

$$\ell_0 = 2 - (1 - \exp(-5t/t_f))$$
$$h_0 = 2 + (1 - \exp(-5t/t_f)).$$

At $t = 0$, the LDL and HDL bloodstream concentrations are $\ell_0 = h = 2$. The LDL concentration decreases with increasing t, possibly in response to a LDL-lowering drug, and eventually at $t = t_f = 1 \times 10^8$, $\ell_0 = 2 - (1 - \exp(-5)) = 1.007$ so that the bloodstream concentration is lowered from 2 to 1.

The HDL concentration increases with increasing t, possibly in response to a HDL-raising drug, and eventually at $t = t_f = 1 \times 10^8$, $h_0 = 2 + (1 - \exp(-5)) = 2.993$ so that the bloodstream concentration is raised from 2 to 3. A HDL/LDL concentration ratio of 3 is usually considered a normal or healthy condition.

ncase=1,2,3 are programmed in the following routines.

4.2   PDE MODEL ROUTINES

The main program is next followed by the ODE/MOL routine.

4.2.1   MAIN PROGRAM

The main program of Listing 2.1 (with $t_f = 1 \times 10^8$) is extended with the specification of ncase.

Listing 4.1: Main program for Eqs. (1.1), (1.2), and (1.3), ncase=1,2,3

```
#
# Six PDE atherosclerosis model
#
# Delete previous workspaces
  rm(list=ls(all=TRUE))
#
# Access ODE integrator
  library("deSolve");
#
# Access functions for numerical solution
  setwd("f:/atherosclerosis/chap4");
  source("pde1a.R");
#
# Select case
  ncase=1;
#
# Select plotting
#
#   ip=1: 2D, vs x, t parametrically
#
#   ip=2: 2D vs t, x=xl
#
#   ip=3: 3D
#
  ip=3;
#
# Parameters
```

```
#
# Diffusivities
  D_l=1.0e-08;  D_h=1.0e-08;
  D_p=1.0e-08;  D_q=1.0e-08;
  D_m=1.0e-08;  D_N=1.0e-08;
#
# Mass transfer coefficients
  k_l=1.0e-08;  k_h=1.0e-08;
  k_p=0.0e-08;  k_q=1.0e-08;
  k_m=0.0e-08;  k_N=0.0e-08;
#
# Bloodstream (lumen) concentrations
  l0=2;  h0=2;
  p0=0;  q0=0;
  m0=0;  N0=0;
#
# Reaction kinetic rate constants
#
# l(x,t)
  r_l1=5.0e-08;  r_l2=5.0e-08;
  r_l3=1.0e-05;
#
# h(x,t);
  r_h1=5.0e-08;  r_h2=5.0e-08;
  r_h3=1.0e-05;
#
# p(x,t)
  r_p1=5.0e-08;  r_p2=5.0e-08;
#
# q(x,t)
  r_q1=5.0e-08;  r_q2=5.0e-08;
#
# m(x,t)
  r_m1=5.0e-08;  r_m2=5.0e-08;
  r_m3=5.0e-08;  r_m4=5.0e-08;
  r_m5=5.0e-08;
#
# N(x,t)
  r_N1=5.0e-08;  r_N2=5.0e-08;
```

```
  r_N3=5.0e-08;
#
# Spatial grid (in x)
  nx=26;
  xl=0;xu=0.004;
  x=seq(from=xl,to=xu,(xu-xl)/(nx-1));
#
# Independent variable for ODE integration
  t0=0;tf=1.0e+08;nout=51;
  tout=seq(from=t0,to=tf,by=(tf-t0)/(nout-1));
#
# Initial condition (t=0)
  u0=rep(0,6*nx);
  for(i in 1:nx){
    u0[i]      =0;
    u0[i+nx]   =0;
    u0[i+2*nx]=0;
    u0[i+3*nx]=0;
    u0[i+4*nx]=0;
    u0[i+5*nx]=0;
  }
  ncall=0;
#
# ODE integration
  out=lsodes(y=u0,times=tout,func=pde1a,
      sparsetype="sparseint",rtol=1e-6,
      atol=1e-6,maxord=5);
  nrow(out)
  ncol(out)
#
# Arrays for plotting numerical solution
  l=matrix(0,nrow=nx,ncol=nout);
  h=matrix(0,nrow=nx,ncol=nout);
  p=matrix(0,nrow=nx,ncol=nout);
  q=matrix(0,nrow=nx,ncol=nout);
  m=matrix(0,nrow=nx,ncol=nout);
  N=matrix(0,nrow=nx,ncol=nout);
  for(it in 1:nout){
    for(i in 1:nx){
```

```
      l[i,it]=out[it,i+1];
      h[i,it]=out[it,i+1+nx];
      p[i,it]=out[it,i+1+2*nx];
      q[i,it]=out[it,i+1+3*nx];
      m[i,it]=out[it,i+1+4*nx];
      N[i,it]=out[it,i+1+5*nx];
    }
  }
#
# Display numerical solution
  iv=seq(from=1,to=nout,by=25);
  for(it in iv){
    cat(sprintf(
      "\n                        t                x"));
    cat(sprintf(
      "\n                l(x,t)          h(x,t)"));
    cat(sprintf(
      "\n                p(x,t)          q(x,t)"));
    cat(sprintf(
      "\n                m(x,t)          N(x,t)"));
  iv=seq(from=1,to=nx,by=25);
  for(i in iv){
    cat(sprintf("\n          %12.2e %12.2e",
                tout[it],x[i]));
    cat(sprintf("\n          %12.3e %12.3e",
                l[i,it],h[i,it]));
    cat(sprintf("\n          %12.3e %12.3e",
                p[i,it],q[i,it]));
    cat(sprintf("\n          %12.3e %12.3e\n",
                m[i,it],N[i,it]));
    }
  }
#
# Calls to ODE routine
  cat(sprintf("\n\n ncall = %5d\n\n",ncall));
#
# Plot PDE solutions
#
# l,h,p,q,m,N against x, parametrically in t
```

```
#
   if(ip==1){
#
# l(x,t)
   par(mfrow=c(1,1));
   matplot(x,l,type="l",xlab="x",ylab="l(x,t)",
     lty=1,main="",lwd=2,col="black");
#
# h(x,t)
   par(mfrow=c(1,1));
   matplot(x,h,type="l",xlab="x",ylab="h(x,t)",
     lty=1,main="",lwd=2,col="black");
#
# p(x,t)
   par(mfrow=c(1,1));
   matplot(x,p,type="l",xlab="x",ylab="p(x,t)",
     lty=1,main="",lwd=2,col="black");
#
# q(x,t)
   par(mfrow=c(1,1));
   matplot(x,q,type="l",xlab="x",ylab="q(x,t)",
     lty=1,main="",lwd=2,col="black");
#
# m(x,t)
   par(mfrow=c(1,1));
   matplot(x,m,type="l",xlab="x",ylab="m(x,t)",
     lty=1,main="",lwd=2,col="black");
#
# N(x,t)
   par(mfrow=c(1,1));
   matplot(x,N,type="l",xlab="x",ylab="N(x,t)",
     lty=1,main="",lwd=2,col="black");
   }
#
# l,h,p,q,m,N against t, x = xl
#
   if(ip==2){
#
# l(x,t)
```

```
  par(mfrow=c(1,1));
  matplot(tout,l[1,],type="l",xlab="t",
    ylab="l(x,t)",lty=1,main="",lwd=2,
    col="black");
#
# h(x,t)
  par(mfrow=c(1,1));
  matplot(tout,h[1,],type="l",xlab="t",
    ylab="h(x,t)",lty=1,main="",lwd=2,
    col="black");
#
# p(x,t)
  par(mfrow=c(1,1));
  matplot(tout,p[1,],type="l",xlab="t",
    ylab="p(x,t)",lty=1,main="",lwd=2,
    col="black");
#
# q(x,t)
  par(mfrow=c(1,1));
  matplot(tout,q[1,],type="l",xlab="t",
    ylab="q(x,t)",lty=1,main="",lwd=2,
    col="black");
#
# m(x,t)
  par(mfrow=c(1,1));
  matplot(tout,m[1,],type="l",xlab="t",
    ylab="m(x,t)",lty=1,main="",lwd=2,
    col="black");
#
# N(x,t)
  par(mfrow=c(1,1));
  matplot(tout,N[1,],type="l",xlab="t",
    ylab="N(x,t)",lty=1,main="",lwd=2,
    col="black");
  }
#
# l,h,p,q,m,N,  3D
#
  if(ip==3){
```

```
  t=tout
#
# l(x,t)
  persp(x,t,l,theta=60,phi=45,
        xlim=c(xl,xu),ylim=c(t0,tf),xlab="x",
        ylab="t",zlab="l(x,t)");
#
# h(x,t)
  persp(x,t,h,theta=60,phi=45,
        xlim=c(xl,xu),ylim=c(t0,tf),xlab="x",
        ylab="t",zlab="h(x,t)");
#
# p(x,t)
  persp(x,t,p,theta=60,phi=45,
        xlim=c(xl,xu),ylim=c(t0,tf),xlab="x",
        ylab="t",zlab="p(x,t)");
#
# q(x,t)
  persp(x,t,q,theta=60,phi=45,
        xlim=c(xl,xu),ylim=c(t0,tf),xlab="x",
        ylab="t",zlab="q(x,t)");
#
# m(x,t)
  persp(x,t,m,theta=60,phi=45,
        xlim=c(xl,xu),ylim=c(t0,tf),xlab="x",
        ylab="t",zlab="m(x,t)");
#
# N(x,t)
  persp(x,t,N,theta=60,phi=45,
        xlim=c(xl,xu),ylim=c(t0,tf),xlab="x",
        ylab="t",zlab="N(x,t)");
  }
```

The main program of Listing 4.1 is taken from Listing 2.1 with a few details included pertaining to the current application as indicated next.

- The R ODE integrator library deSolve is accessed. Then the directory with the files for the solution of Eqs. (1.1), (1.2), and (1.3) is designated.

```
#
# Access ODE integrator
  library("deSolve");
#
# Access functions for numerical solution
  setwd("f:/atherosclerosis/chap4");
  source("pde1a.R");
```

pde1a.R is the routine for the method of lines (MOL) approximation of PDEs (1.1) (discussed subsequently).

- ncase is specified, starting with ncase=1 as a base case, as explained in the preceding introduction.

```
#
# Select case
  ncase=1;
```

- The time interval is $0 \le t \le 1 \times 10^8$.

```
  t0=0;tf=1.0e+08;nout=51;
```

- pde1a is called by lsodes for the integration of the $6(26) = 156$ MOL/ODEs.

```
#
# ODE integration
  out=lsodes(y=u0,times=tout,func=pde1a,
      sparsetype="sparseint",rtol=1e-6,
      atol=1e-6,maxord=5);
  nrow(out)
  ncol(out)
```

pde1a is discussed next.

4.2.2   ODE/MOL ROUTINE

pde1a is taken from Listing 2.2 with coding added for ncase=1,2,3 as explained in the preceding introduction (Section 4.1).

Listing 4.2: ODE/MOL routine pde1a for Eqs. (1.1), (1.2), (1.3) with ncase=1,2,3

```
pde1a=function(t,u,parms){
#
# Function pde1a computes the t derivative
# vectors of l(x,t),h(x,t),p(x,t),q(x,t),
# m(x,t),N(x,t)
#
# One vector to six vectors
  l=rep(0,nx);h=rep(0,nx);
  p=rep(0,nx);q=rep(0,nx);
  m=rep(0,nx);N=rep(0,nx);
  for(i in 1:nx){
    l[i]=u[i];
    h[i]=u[i+nx];
    p[i]=u[i+2*nx];
    q[i]=u[i+3*nx];
    m[i]=u[i+4*nx];
    N[i]=u[i+5*nx];
  }
#
# lx,hx,px,qx,mx,Nx
  lx=rep(0,nx);hx=rep(0,nx);
  px=rep(0,nx);qx=rep(0,nx);
  mx=rep(0,nx);Nx=rep(0,nx);
  tablel=splinefun(x,l);lx=tablel(x,deriv=1);
  tableh=splinefun(x,h);hx=tableh(x,deriv=1);
  tablep=splinefun(x,p);px=tablep(x,deriv=1);
  tableq=splinefun(x,q);qx=tableq(x,deriv=1);
  tablem=splinefun(x,m);mx=tablem(x,deriv=1);
  tableN=splinefun(x,N);Nx=tableN(x,deriv=1);
#
# BCs
  if(ncase==1){l0=2;h0=2;}
  if(ncase==2){l0=2-(1-exp(-5*t/tf));
               h0=2;}
  if(ncase==3){l0=2-(1-exp(-5*t/tf));
               h0=2+(1-exp(-5*t/tf));}
  lx[1]=-(k_1/D_1)*(l0-l[1]);lx[nx]=0;
```

```
    hx[1]=-(k_h/D_h)*(h0-h[1]);hx[nx]=0;
    px[1]=-(k_p/D_p)*(p0-p[1]);px[nx]=0;
    qx[1]=-(k_q/D_q)*(q0-q[1]);qx[nx]=0;
    mx[1]=-(k_m/D_m)*(m0-m[1]);mx[nx]=0;
    Nx[1]=-(k_N/D_N)*(N0-N[1]);Nx[nx]=0;
#
#   lxx,hxx,pxx,qxx,mxx,Nxx
    lxx=rep(0,nx);hxx=rep(0,nx);
    pxx=rep(0,nx);qxx=rep(0,nx);
    mxx=rep(0,nx);Nxx=rep(0,nx);
    tablelx=splinefun(x,lx);lxx=tablelx(x,deriv=1);
    tablehx=splinefun(x,hx);hxx=tablehx(x,deriv=1);
    tablepx=splinefun(x,px);pxx=tablepx(x,deriv=1);
    tableqx=splinefun(x,qx);qxx=tableqx(x,deriv=1);
    tablemx=splinefun(x,mx);mxx=tablemx(x,deriv=1);
    tableNx=splinefun(x,Nx);Nxx=tableNx(x,deriv=1);
#
#   PDEs
    lt=rep(0,nx);ht=rep(0,nx);
    pt=rep(0,nx);qt=rep(0,nx);
    mt=rep(0,nx);Nt=rep(0,nx);
    for(i in 1:nx){
#
#     Product functions
      lm=l[i]/(r_l3+l[i])*m[i];
      hN=h[i]/(r_h3+h[i])*N[i];
#
#     l_t
      lt[i]=D_l*lxx[i]-
        r_l1*lm-r_l2*l[i];
#
#     h_t
      ht[i]=D_h*hxx[i]-
        r_h1*hN-r_h2*h[i];
#
#     p_t
      pt[i]=D_p*pxx[i]+
        r_p1*lm-r_p2*p[i];
#
```

```
#    q_t
     qt[i]=D_q*qxx[i]+
       r_q1*lm-r_q2*q[i];
#
#    m_t
     mt[i]=D_m*mxx[i]-
       r_m1*(mx[i]*lx[i]+m[i]*lxx[i])-
       r_m2*lm+r_m3*hN-r_m4*m[i]+
       r_m5*l[i];
#
#    N_t
     Nt[i]=D_N*Nxx[i]+
       r_N1*lm-r_N2*hN+
       r_N3*l[i];
   }
#
# Six vectors to one vector
   ut=rep(0,6*nx);
   for(i in 1:nx){
     ut[i]      =lt[i];
     ut[i+nx]   =ht[i];
     ut[i+2*nx]=pt[i];
     ut[i+3*nx]=qt[i];
     ut[i+4*nx]=mt[i];
     ut[i+5*nx]=Nt[i];
   }
#
# Increment calls to pde1a
   ncall <<- ncall+1;
#
# Return derivative vector
   return(list(c(ut)));
   }
```

Listing 4.2 is taken from Listing 2.2. The only change in Listing 4.2 is additional coding to define ℓ_0, h_0 in BCs (1.2-1) and (1.2-3) numerically for ncase=1,2,3.

```
#
# BCs
  if(ncase==1){10=2;h0=2;}
  if(ncase==2){10=2-(1-exp(-5*t/tf));
                h0=2;}
  if(ncase==3){10=2-(1-exp(-5*t/tf));
                h0=2+(1-exp(-5*t/tf));}
  lx[1]=-(k_1/D_1)*(10-1[1]);lx[nx]=0;
  hx[1]=-(k_h/D_h)*(h0-h[1]);hx[nx]=0;
  px[1]=-(k_p/D_p)*(p0-p[1]);px[nx]=0;
  qx[1]=-(k_q/D_q)*(q0-q[1]);qx[nx]=0;
  mx[1]=-(k_m/D_m)*(m0-m[1]);mx[nx]=0;
  Nx[1]=-(k_N/D_N)*(N0-N[1]);Nx[nx]=0;
```

The numerical and graphical output for the main program and pde1a follows.

4.2.3   MODEL OUTPUT

We can note the following details about the output in Table 4.1.

- The dimensions of the solution matrix out are $nout \times 6nx + 1 = 51 \times 6(26) + 1 = 157$.

  [1] 51

  [1] 157

  The offset $+1$ results from the value of t as the first element in each of the $nout = 51$ solution vectors. These same values of t are in tout,

- ICs (1.3) $(t = 0)$ are verified for the homogeneous (zero) case.

- The output is for $x = 0, 0.004/25, ..., 0.004$ as programmed in Listing 4.1 (26 values of x at each value of t with every 25th value in x displayed so $x = 0, 0.004$).

- The output is for $t = 0, 1 \times 10^8/50, ..., 1 \times 10^8$ as programmed in Listing 4.1 (51 values of t with every 25th value in t displayed).

- The solutions $\ell(x, t)$ to $N(x, t)$ approach a steady state (equilibrium) solution at $t = t_f = 1 \times 10^8$ (this is clarified by considering the graphical output of Figs. 4.1 next). In particular, $\ell(x = 0, t = 1 \times 10^8) = $ 1.955e+00, $h(x = 0, t = 1 \times 10^8) = $ 1.921e+00 which are close to $\ell_0 = h_0 = 2$ of BCs (1.2-1) and (1.2-3) for ncase=1.

- The variation of the solutions with respect to x is small, a consequence of the thin endothelial inner layer (EIL), which is also apparent in Figs. 4.1.

Table 4.1: Numerical output for ncase=1 *(Continues.)*

[1] 51

[1] 157

l(x,t)	h(x,t)
p(x,t)	q(x,t)
m(x,t)	N(x,t)
0.00e+00	0.00e+00
0.000e+00	0.000e+00
0.000e+00	0.000e+00
0.000e+00	0.000e+00
0.00e+00	4.00e-03
0.000e+00	0.000e+00
0.000e+00	0.000e+00
0.000e+00	0.000e+00

t	x
l(x,t)	h(x,t)
p(x,t)	q(x,t)
m(x,t)	N(x,t)
5.00e+07	0.00e+00
1.955e+00	1.921e+00
2.504e-01	5.876e-03
3.000e-01	2.042e+00
5.00e+07	4.00e-03
1.955e+00	1.921e+00
2.504e-01	5.888e-03
3.000e-01	2.042e+00

Table 4.1: *(Continued.)* Numerical output for `ncase=1`

t	x
l(x,t)	h(x,t)
p(x,t)	q(x,t)
m(x,t)	N(x,t)
1.00e+08	0.00e+00
1.955e+00	1.917e+00
3.073e-01	6.172e-03
3.148e-01	2.249e+00
1.00e+08	4.00e-03
1.955e+00	1.917e+00
3.073e-01	6.184e-03
3.148e-01	2.249e+00

```
ncall =   1030
```

- The computational effort is acceptable, `ncall = 1030`.

The graphical output of Figs. 4.1 follows.

In Figs. 4.1-1 and 4.1-2, $\ell(x,t)$, $h(x,t)$ respond rapidly to $\ell_0 = h_0 = 2$ and reach a steady state soon after $t = 0$. This rapid response is due to the effect of BCs (1.2-1) and (1.2-3).

Figures 4.1 indicate that the six PDE dependent variables $\ell(x,t)$ to $N(x,t)$ approach steady state values listed in Table 4.1. In particular, $N(x,t)$ moves from the homogeneous IC (1.3-6) to a steady state value `2.249e+00` (Table 4.1) indicating a significant increase in the foam cell density, and as a consequence, possible arterial hardening and plaque buildup (implicated in stroke and heart attacks).

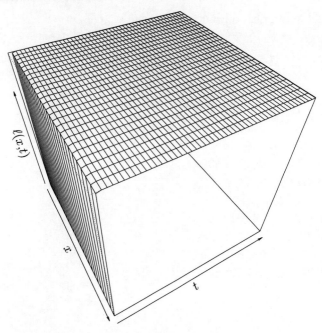

Figure 4.1-1: Numerical solution $\ell(x, t)$ from Eq. (1.1-1), ncase=1.

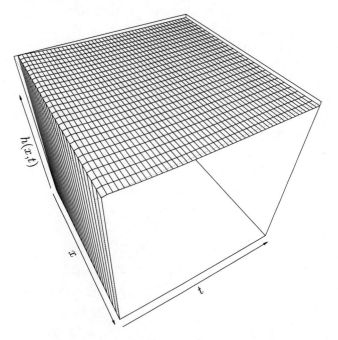

Figure 4.1-2: Numerical solution $h(x, t)$ from Eq. (1.1-2), ncase=1.

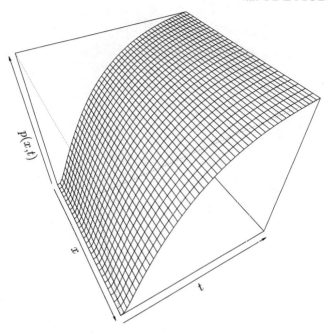

Figure 4.1-3: Numerical solution $p(x,t)$ from Eq. (1.1-3), ncase=1.

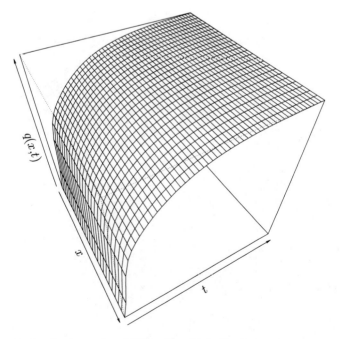

Figure 4.1-4: Numerical solution $q(x,t)$ from Eq. (1.1-4), ncase=1.

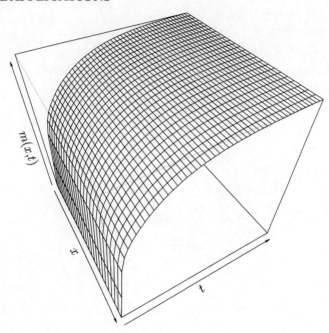

Figure 4.1-5: Numerical solution $m(x,t)$ from Eq. (1.1-5), `ncase=1`.

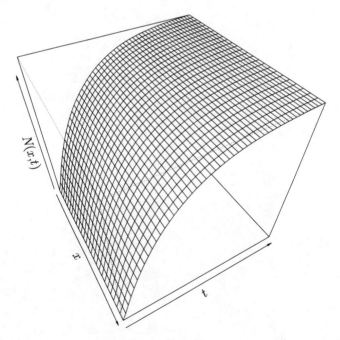

Figure 4.1-6: Numerical solution $N(x,t)$ from Eq. (1.1-6), `ncase=1`.

This base case `ncase=1` solution can now be used for comparison with the solutions for `ncase=2,3` considred next. The output for `ncase=2` (Listing 4.1) appears in Table 4.2.

We can note the following details in Table 4.2.

- The modified LDL concentration, $\ell(x,t)$, is sigificantly reduced for `ncase=2`. For example, at $t = 1 \times 10^8$, $x = 0$,

```
ncase=1 (Table 4.1)
                 t               x
            l(x,t)          h(x,t)
          1.00e+08        0.00e+00
          1.955e+00       1.917e+00


ncase=2 (Table 4.2)
                 t               x
            l(x,t)          h(x,t)
          1.00e+08        0.00e+00
          9.817e-01       1.936e+00
```

For `ncase=2`, $\ell(x,t)$ approaches the bloodstream value $\ell_0 = 1$. $h(x,t)$ is only slightly affected.

- Similarly, the foam cell density $N(x,t)$ is significantly reduced.

```
ncase=1 (Table 4.1)
                 t               x
            m(x,t)          N(x,t)
          1.00e+08        0.00e+00
          3.148e-01       2.249e+00


ncase=2 (Table 4.2)
                 t               x
            m(x,t)          N(x,t)
          1.00e+08        0.00e+00
          2.745e-01       1.279e+00
```

This result is noteworthy since the lower LDL in the bloodstream reduces the foam cell density (and thereby, arterial hardening and plaque formation).

Table 4.2: Numerical output for `ncase=2` *(Continues.)*

[1] 51

[1] 157

t	x
l(x,t)	h(x,t)
p(x,t)	q(x,t)
m(x,t)	N(x,t)
0.00e+00	0.00e+00
0.000e+00	0.000e+00
0.000e+00	0.000e+00
0.000e+00	0.000e+00
0.00e+00	4.00e-03
0.000e+00	0.000e+00
0.000e+00	0.000e+00
0.000e+00	0.000e+00

t	x
l(x,t)	h(x,t)
p(x,t)	q(x,t)
m(x,t)	N(x,t)
5.00e+07	0.00e+00
1.057e+00	1.934e+00
2.645e-01	5.717e-03
2.914e-01	1.361e+00
5.00e+07	4.00e-03
1.057e+00	1.934e+00
2.645e-01	5.728e-03
2.914e-01	1.361e+00

Table 4.2: *(Continued.)* Numerical output for `ncase=2`

t	x
l(x,t)	h(x,t)
p(x,t)	q(x,t)
m(x,t)	N(x,t)
1.00e+08	0.00e+00
9.817e-01	1.936e+00
2.772e-01	5.383e-03
2.745e-01	1.279e+00
1.00e+08	4.00e-03
9.817e-01	1.936e+00
2.772e-01	5.395e-03
2.745e-01	1.279e+00

```
ncall =   1047
```

Figure 4.2-1 indicates $\ell(x,t)$ goes through a maximum, then begins to decline. Figure 4.2-6 indicates that $N(x,t)$ also eventually declines.

The 3D graphical output for $h(x,t)$ to $m(x,t)$ is not included to conserve space. In conclusion, for `ncase=2` the bloodstream lowering (possibly in response to a therapeutic drug) reduces the foam cell density $N(x,t)$ in the EIL. The output for `ncase=3` appears in Table 4.3. We can note the following details in Table 4.3.

- The HDL concentration, $h(x,t)$, is sigificantly increased for `ncase=3`. For example, at $t = 1 \times 10^8$, $x = 0$,

```
ncase=1 (Table 4.1)
                t               x
        l(x,t)          h(x,t)
        1.00e+08        0.00e+00
        1.955e+00       1.917e+00


ncase=3 (Table 4.3)
                t               x
        l(x,t)          h(x,t)
        1.00e+08        0.00e+00
        9.817e-01       2.909e+00
```

For `ncase=3`, $\ell(x,t)$, and $h(x,t)$ approach the bloodstream values $\ell_0 = 1$ and $h_0 = 3$.

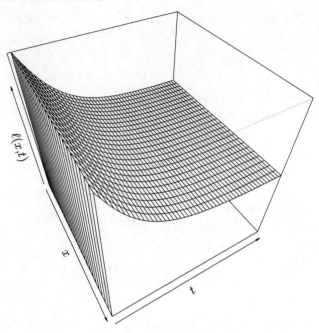

Figure 4.2-1: Numerical solution $\ell(x,t)$ from Eq. (1.1-1), ncase=2.

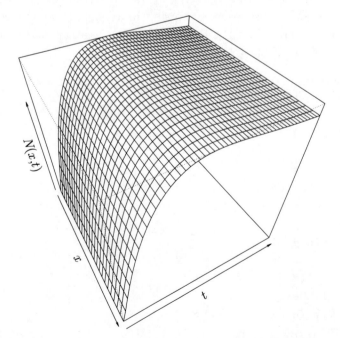

Figure 4.2-6: Numerical solution $N(x,t)$ from Eq. (1.1-6), ncase=2.

Table 4.3: Numerical output for ncase=3 *(Continues.)*

[1] 51

[1] 157

```
                t                x
           l(x,t)           h(x,t)
           p(x,t)           q(x,t)
           m(x,t)           N(x,t)
         0.00e+00         0.00e+00
        0.000e+00        0.000e+00
        0.000e+00        0.000e+00
        0.000e+00        0.000e+00

         0.00e+00         4.00e-03
        0.000e+00        0.000e+00
        0.000e+00        0.000e+00
        0.000e+00        0.000e+00

                t                x
           l(x,t)           h(x,t)
           p(x,t)           q(x,t)
           m(x,t)           N(x,t)
         5.00e+07         0.00e+00
        1.057e+00        2.832e+00
        2.645e-01        5.717e-03
        2.914e-01        1.361e+00

         5.00e+07         4.00e-03
        1.057e+00        2.832e+00
        2.645e-01        5.728e-03
        2.914e-01        1.361e+00
```

Table 4.3: *(Continued.)* Numerical output for ncase=3

```
            t              x
        l(x,t)         h(x,t)
        p(x,t)         q(x,t)
        m(x,t)         N(x,t)
       1.00e+08       0.00e+00
      9.817e-01       2.909e+00
      2.772e-01       5.383e-03
      2.745e-01       1.279e+00


       1.00e+08       4.00e-03
      9.817e-01       2.909e+00
      2.772e-01       5.394e-03
      2.745e-01       1.279e+00


ncall =   1050
```

- Also, for ncase=3, the foam cell density $N(x, t)$ for ncase=1 is significantly reduced, but there is no effect on $N(x, t)$ for ncase=2.

```
ncase=1 (Table 4.1)
                t              x
            m(x,t)         N(x,t)
           1.00e+08       0.00e+00
          3.148e-01       2.249e+00


ncase=2 (Table 4.2)
                t              x
            m(x,t)         N(x,t)
           1.00e+08       0.00e+00
          2.745e-01       1.279e+00


ncase=3 (Table 4.3)
                t              x
            m(x,t)         N(x,t)
           1.00e+08       0.00e+00
          2.745e-01       1.279e+00
```

These results suggest that a HDL-raising drug will have no appreciable effect on the foam cell density $N(x, t)$. This result can be studied by comparing the RHS terms of Eq. (1.1-6) and the coding in pde1a of Listing 4.2. Specifically, the term pertaining to $h(x,t)$ in Eq. (1.1-6)

$$-r_{N2}\frac{h}{r_{h3} + h}N$$

appears to have little effect relative to the terms pertaining to $\ell(x,t)$

$$r_{N1}\frac{\ell}{r_{l3} + \ell}m$$

$$r_{N3}\ell.$$

This conclusion can be confirmed by computing and displaying these terms as illustrated in Chapter 3. If this additional analysis explains why increasing HDL in ncase=3 has no effect on $N(x, t)$, then additional analysis with variation in the rate constants r_{N1}, r_{N2}, and r_{N3} would be of interest, particularly if they can be estimated for proposed HDL-raising drugs. This extended analysis is left as an exercise.

To conclude this discussion of the solution for ncase=3, selected graphical output of Figs. 4.3 follows.

Figure 4.3-1 is similar to Fig. 4.2-1 for ncase=2 and reflects the reduced modified LDL, $\ell(x,t)$.

Figure 4.3-2 reflects the increased $h(x,t)$ for ncase=1 (in Fig. 4.1-2).

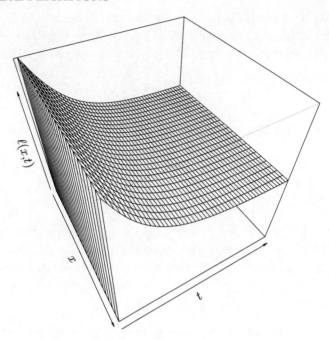

Figure 4.3-1: Numerical solution $\ell(x,t)$ from Eq. (1.1-1), ncase=3.

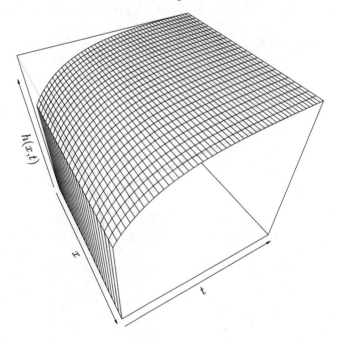

Figure 4.3-2: Numerical solution $h(x,t)$ from Eq. (1.1-2), ncase=3.

Figure 4.3-6 is similar to Fig. 4.2-6 for `ncase=2`, that is, no effect on $N(x,t)$ of the increased HDL for `ncase=3`.

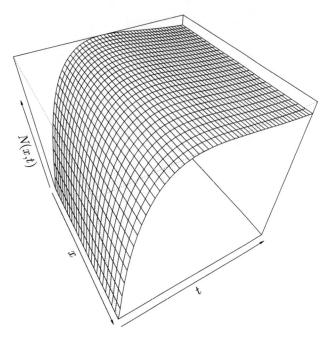

Figure 4.3-6: Numerical solution $N(x,t)$ from Eq. (1.1-6), `ncase=3`.

4.3   SUMMARY AND CONCLUSIONS

The examples in this chapter suggest a methodology for the investigation of the effect of LDL-lowering and HDL-raising therapeutic drugs. The examples are based on specific values of the parameters defined in the main program of Listing 4.1. The dependency of the features (characteristics) of the model solutions on the specific parameter values emphasizes the importance of parameter estimation in the development and use of computer-based PDE models.

REFERENCES

[1] Linsel-Nitschke, P. and Tall, A. R. (2005), HDL as a target in the treatment of atherosclerotic cardiovascular disease, *Nature Reviews Drug Discovery*, 4, pp. 193–205. DOI: 10.1038/nrd1658. 97

Author's Biography

WILLIAM E. SCHIESSER

William E. Schiesser is Emeritus McCann Professor of Computational Biomedical Engineering and Chemical and Biomolecular Engineering at Lehigh University.

Index

Printed in the United States
by Baker & Taylor Publisher Services